ISBN 978-1-5282-2523-6
PIBN 10899113

This book is a reproduction of an important historical work. Forgotten Books uses
state-of-the-art technology to digitally reconstruct the work, preserving the original format
whilst repairing imperfections present in the aged copy. In rare cases, an imperfection in
the original, such as a blemish or missing page, may be replicated in our edition. We do,
however, repair the vast majority of imperfections successfully; any imperfections that
remain are intentionally left to preserve the state of such historical works.

For support please visit www.forgottenbooks.com

Historic, archived document

Do not assume content reflects current
scientific knowledge, policies, or practices.

CALIBRATION OF SELECTED
INFILTRATION EQUATIONS
FOR THE GEORGIA COASTAL PLAIN

ARS-S-113
July 1976

AGRICULTURAL RESEARCH SERVICE • U.S. DEPARTMENT OF AGRICULTURE

Agricultural Research Service
UNITED STATES DEPARTMENT OF AGRICULTURE
in cooperation with
University of Georgia College of Agriculture Experiment Stations
Georgia Institute of Technology
and
Middle South Georgia Soil Conservation District

CONTENTS

ILLUSTRATIONS

TABLES

ACKNOWLEDGMENTS

The authors gratefully acknowledge the assistance of Russell R. Bruce, soil scientist, and Albert D. Lovell, agricultural research technician, both of the Southern Piedmont Conservation Research Center, Agricultural Research Service (ARS), U.S. Department of Agriculture (USDA), Watkinsville, Ga. Bruce provided valuable guidance in site selection and program planning and loaned us an infiltrometer and associated equipment. Lovell furnished very helpful technical guidance in setting up and operating the infiltrometer. We also wish to thank Willard M. Snyder, Southeast Watershed Research Center, ARS, USDA, Athens, Ga., for his assistance in site selection and program planning. The authors are also indebted to John W. Calhoun, Soil Conservation Service, Tifton, Ga., for assistance in site selection and soil classification.

ii

CALIBRATION OF SELECTED INFILTRATION EQUATIONS FOR THE GEORGIA COASTAL PLAIN

By Walter Rawls,[1] Paul Yates,[2] and Loris Asmussen[3]

ABSTRACT

Experimental infiltration data were obtained for 11 Coastal Plain soils. A total of 77 infiltration runs were made. Initial and final soil-moisture measurements, detailed soil-profile descriptions, and moisture-tension measurements were also made. The final infiltration rates ranged between 0.12 in/h and 4.61 in/h (0.30–11.71 cm/h) for all soils except Kershaw. At a rainfall application rate of 6.25 in/h (15.88 cm/h) all water continued to infiltrate on Kershaw soils.

The data were fit to equations proposed by Horton, Green and Ampt, Phillip, Holtan, and Snyder. Adequate determinations of infiltration rates were obtained with each of the five equations tested, but the best representations of the infiltration-capacity curves were obtained from Horton's and Snyder's equations. (Snyder's equation is capable of explaining recovery during periods when rainfall rate falls below infiltration capacity, while Horton's is not.) Green and Ampt's, Phillip's, and Holtan's equations consistently overestimated the early portion of the infiltration-capacity curve and underestimated the later portion. The wide variation in equation parameters between applications in situations in which similar initial conditions existed on the same soil was caused by experimental error rather than by the fit of the equations to the data. The results of the fittings can nonetheless be used as a guide for applying the equations in the Coastal Plain.

INTRODUCTION

Water infiltration data for different soils are essential for good land-use planning and as aids in coming to a more thorough understanding of the rainfall-runoff process. Infiltration data for agricultural soils of the Coastal Plain of the Southeastern United States are virtually non-existent. Therefore, an exploratory study was conducted to obtain infiltration data and to examine the infiltration process in selected soils. Objectives of the study were (1) to obtain infiltration data on selected soils in the Coastal Plain, (2) to obtain supplementary data to further understanding of the infiltration process, and (3) to calibrate existing infiltration equations for the Coastal Plain.

In general, the problem of characterizing infiltration is one of describing the flow of water through porous media. Darcy's law and the law of conservation of mass may be used to derive a general equation of flow,

$$\frac{\partial \Theta}{\partial t} = \frac{\partial}{\partial x}\left(K\frac{\partial h}{\partial x}\right) + \frac{\partial K}{\partial x}, \tag{1}$$

where Θ = moisture content (percent),

[1] Hydrologist, Hydrograph Laboratory, Plant Physiology Institute, Agricultural Research Service, U.S. Department of Agriculture, Beltsville, Md. 20705.

[2] Hydraulic engineer, Southeast Watershed Research Laboratory, Agricultural Research Service, U.S. Department of Agriculture, Athens, Ga. 30601.

[3] Geologist, Southeast Watershed Research Unit, Agricultural Research Service, U.S. Department of Agriculture, Tifton, Ga. 31794.

t=time (minutes),

x=position (feet),

h=capillary potential (feet),

and K=hydraulic conductivity, a function of moisture content (feet per minute).

Recently there has been an increase in research dealing with characterization of porous media flow and development of numerical techniques for solution of equation 1 under different boundary conditions. Amerman (1)[4] has discussed advances in modern infiltration theory. Despite recent progress, however, numerical techniques for the solution of equation 1 and the characterization of infiltration do not yet produce satisfactory results when applied to field-scale problems.

A number of algebraic equations have been proposed to determine field-scale infiltration rates. The equations presented by Horton (9), Green and Ampt (6), Phillip (14), Holtan (7), Holtan et al. (8), and Snyder (20) were chosen for evaluation in this study. There have been attempts to evaluate these equations on the basis of experimental data and to obtain numerical values for the parameters in areas other than the Coastal Plain (17). Because little is known of the equation parameters for Coastal Plain soils, however, it has been difficult to use the equations in this area.

Horton (9) presented one of the most widely used infiltration equations,

$$f = f_c + (f_0 - f_c) \, e^{K_f t}, \qquad (2)$$

where f=infiltration capacity at time t (inches per hour),

f_c=final constant infiltration capacity as $t \to \infty$ (inches per hour),

f_0=infiltration capacity at t=0 (inches per hour),

e=base of Napierian logarithms,

K_f=constant governing the rate of change of infiltration capacity with time,

and t=time from the beginning of rain or beginning of runoff (minutes per hour).

Horton showed that equation 2 may be derived rationally from the simple assumption that reduction in infiltration capacity during rain is

---

[4] Italic numbers in parentheses refer to items in "Literature Cited" preceding the appendix.

$$f = a(S_t - F)^n + f_c, \qquad (6)$$

where f=infiltration capacity (inches per hour),

a and n=constants dependent on the soil type, surface, and cropping conditions,

S_t=storage potential of a soil above the impeding layer [total porosity minus the antecedent soil moisture (inches)],

F=accumulated infiltration (inches),

and f_c=steady state infiltration rate (inches per hour).

Overton (11) thoroughly discusses the above equation. Using soil moisture instead of time as the independent variable makes it possible to compute the infiltration capacity at any time during a storm, even when rainfall does not exceed the infiltration capacity, or when there is a temporary interruption in rainfall.

Snyder (20) presented a watershed retention function, based on macroscale concept in the watershed physical process, which takes the form

$$f_t = f_{t-1} - (a + bf_{t-1}) \frac{(R_t + f_a - f_{t-1})}{R_t + f_a - f_c} \frac{(R_t - f_c)}{R_t + f_c}$$
$$(f_{t-1} - f_c)\Delta t, \qquad (7)$$

where f_t=infiltration capacity at time t (inches per hour),

f_{t-1}=infiltration capacity at time $t - \Delta t$ (inches per hour),

a and b=shape constants,

R_t=rainfall during time Δt (inches),

f_a=upper dry limit of infiltration (inches per hour),

f_c=lower limit of infiltration at saturation (inches per hour),

and Δt=time increment (minutes).

Even though equation 7 was not proposed as an infiltration equation, it was an acceptable substitute and worthy of study. Equation 7 has the capacity to explain infiltration recovery during periods of no rainfall or low intensity rainfall, which capacity is necessary if an infiltration equation is to be applicable to a watershed. Smith (18) has discussed this equation.

STUDY AREA

The general area chosen for study was in the Tifton Upland Physiographic Region of the Georgia Coastal Plain (fig. 1). This region represents 34.4% of the Coastal Plain.

The specific study area was in south-central Georgia, near Tifton, and in the vicinity of the Little River Experimental Watershed (fig. 1). The surface soils of the watershed exhibit a distinct A and B horizon, with texture ranging from a loamy sand to a sandy clay loam in the top 3 to 4 feet (91 to 122 cm), except for Kershaw, which is a coarse sand. At the 3- to 4-ft depth, there is usually a semipermeable clay layer. Nearly 90% of the area has a slope of 5% or less, but along the sides of some valleys, slope ranges from 5% to 15%.

The Little River Experimental Watershed is composed of 20 soil series, from which 11 were chosen for investigation (23). The 11 soil series account for 95% of the 126-mi² (326-km²) Little River Experimental Watershed, and 91% of the land area in Tift County, Ga. These 11 soil series represent a wide range of physical conditions, as table 1 shows.

Sites were selected from those soils for which soil-profile descriptions (25) and moisture-tension data (10) were available. Within each soil series, accessibility and cover conditions primarily determined specific site selection. Site locations are shown in figure 1.

Because of its latitude and proximity to the warm waters of the Gulf of Mexico and the Atlantic Ocean, the study area experiences long, hot, humid summers and short, mild winters. Average monthly temperatures vary between 52° F (11° C) in January and 81° F (27° C) in July and August, with an average annual temperature of 66° F (19° C). Freezing weather occurs between the middle of November and the middle of March, the most severe freezes occurring in January. The average frost-free season is 253 days.

Precipitation extremes range from a low of 23.25 in (59.1 cm) in 1954 to a high of 70.55 in (179.2 cm) in 1928. The 46.39 in (117.8 cm) measured in 1969 (date of infiltration studies) barely exceeded the 49-year (1922–70) average of 45.77 in (116.3 cm).

The 31-yr (1940–70) average annual pan evaporation is 56.23 in (142.8 cm), the average monthly values varying between 2.21 in (5.6 cm) in December and 7.21 in (18.3 cm) in May.

3

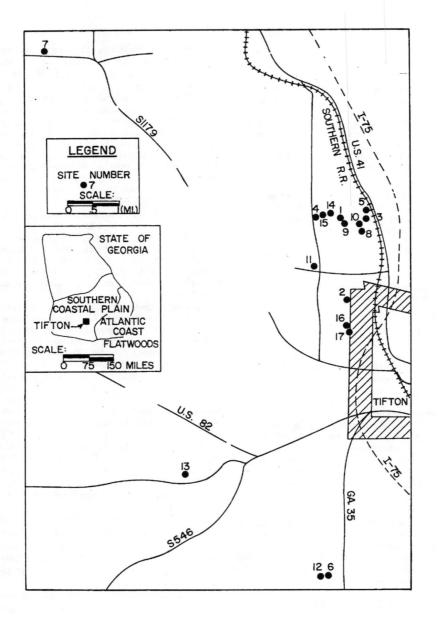

FIGURE 1.—Location of infiltration sites.

TABLE 1.—*Watershed characteristics of the soil series*[1]

Topographic position and soil series	Dominant relief (percent)	Most common use	Percent of land area in Little River Watershed[2]
Upland:			
Cowarts	5–7	Mostly crops; some pasture; little forest.	([2])
Fuquay	0–5	Forest	15.0
Kershaw	0–5	Scrub oak	([2])
Dothan	2–5	Crops	4.6
Middle:			
Troup	2–5	Crops & forest	.3
Carnegie	5–7	Crops & forest	4.5
Tifton	2–5	Crops; pasture; little forest	58.0
Stilson	0–2	Forest; some pasture & crops	2.5
Leefield	0–2	Forest; some pasture & crops	1.5
Lowland:			
Robertsdale	0–2	Forest	([2])
Alapaha	0–2	Forest	9.0

[1] Data from Soil Conservation Service, U.S. Department of Agriculture (*23*).
[2] Less than 0.1% of the total area.

PROCEDURES

INFILTROMETER

The Purdue sprinkling infiltrometer, as developed by Bertrand and Parr (*2*) and modified by Dixon and Peterson (*4*), was used to apply artificial rainfall to the plots. The rainfall was produced from a single full-cone nozzle centered at the top of a 9-ft (2.74-m) tower. The nozzle applied constant rainfall to a circular area of about 100 ft^2 (9.3 m^2). Rainfall rate can be varied from 2.50 in/h to greater than 6 in/h (6.35–15.24 cm/h) by exchanging nozzles. All nozzles produced drop size distributions, final drop velocities, and kinetic energy closely resembling those produced by natural rainfall. Nozzle intensities were checked before each run by applying rainfall for 10 min and collecting the runoff in a calibration pan. To prevent wetting the soil during calibration, the area around the calibration pan was covered with plastic.

A 3.81-ft (1.16-m), square, metal plot frame was centered on the ground beneath the nozzle, and manually driven 2 in (5.1 cm) into the ground, as specified by Bertrand and Parr (*2*). The sides of the plot frame extended 4 in (10.2 cm) above the ground. A covered flume was attached to the downhill side of the frame. Runoff was carried by a vacuum system from the flume into a collection tank where the water stage data

were recorded and converted to rates of runoff. Adjacent to the runoff plot on the upslope side, a neutron probe access tube was installed to measure soil moisture. These tubes, made of 2-in (5.1-cm) o.d. aluminum irrigation tubing, were installed in snug-fitting holes predrilled by a power-driven flight auger. All access tubes were installed at least 1 month prior to time of use.

INFILTRATION FIELD PROCEDURE

The procedure in the paragraphs below was developed to give infiltration data for dry and wet soil conditions. For replication purposes the procedure was performed on two plots at each site. Data shown in table 2 represent results from both plots.

Natural conditions were preserved at each site, except where vegetation cover was excessively tall. Tall vegetation was cut to a height of 6 to 8 in (15.2 to 20.3 cm). Vegetation normally provided between 50% and 80% ground cover. Artificial rainfall was initially applied at a high intensity (4.45 to 6.73 in/h, or 11.30 to 17.09 cm/h) for a time sufficient to cause a relatively constant runoff rate. Rainfall was then stopped for approximately 1 h, after which time it was begun again at a lower intensity (2.64 to 5.17 in/h, or 6.70 to 13.13 cm/h) and continued at this lower intensity until a relatively constant

(Continued on page 8.)

TABLE 2.—*Infiltration data*

Soil series	Hydrologic group[1]	Cover conditions	Identification code[2]	Soil moisture (in) at— 0"-12"	12"-36"	Total length applied rainfall (min)	Time from start of applied rainfall to start of runoff (min)	Rainfall intensity (in/h)	Final infiltration rate (in/h)	Time between applied rainfall (min)
Alapaha loamy sand.	D	Weeds (90%), bare (10%).	01011D	2.30	5.18	140	3	6.73	2.10	85
			01011W	3.36	5.89	120	3	4.45	.72	
Do	D	......do......	01012D	2.33	6.64	130	4	4.57	.44	65
			01012W	3.48	7.11	.20	4	2.88	.31	
Carnegie sandy loam.	C	Grass (100%)	02021D	.93	5.43	135	7	6.61	3.19	60
			02021W	2.67	6.50	123	9	5.53	2.02	
Do	C	......do......	02022D	.96	5.93	150	4	4.69	2.60	65
			02022W	2.76	6.77	110	9	3.00	1.67	
Cowarts loam sand.	C	Weeds (80%), bare (20%).	03031D	1.93	6.93	130	8	5.17	3.03	85
			03031W	3.06	7.38	120	5	6.01	4.61	
			03031WW	3.30	7.11	40	2	5.17	2.99	10
Do	C	......do......	03032D	1.41	6.17	150	30	3.37	3.18	80
			03032W	2.71	6.60	90	6	4.81	1.61	
			03032WW	3.39	6.71	50	8	3.37	2.10	30
Do	C	Weeds (60%), corn (40%).	16031D	2.15	6.47	140	4	4.45	3.48	63
			16031W	2.99	6.80	147	3	6.50	2.81	
Do	C	......do......	16032D	2.35	7.30	150	15	4.45	4.22	85
			16032W	3.35	9.45	135	2	6.50	3.36	
Do	C	......do......	16033D	2.28	7.75	163	7	4.33	4.04	50
			16033W	3.58	8.41	120	4	2.88	.95	
Do	C	Grass (100%)	17031D	1.11	5.85	150	4	4.38	4.07	65
			17031W	2.44	6.47	120	8	6.25	2.13	
Do	C	......do......	17032D	.97	5.53	155	7	4.57	2.63	70
			17032W	2.70	6.61	137	6	6.25	1.05	
Do	C	......do......	17033D	1.02	5.42	130	4	2.76	2.63	60
			17033W	2.24	6.41	100	6	5.29	2.94	
			17033WW	2.76	6.52	70	8	2.76	2.52	25
Dothan loamy sand.	B	Bare (80%), weeds (20%).	04041D	.79	4.04	180	10	4.69	3.79	60
			04041W	2.87	6.43	110	6	3.37	2.71	

Soil		Cover	Sample No.							
Fuquay loamy sand.	B	Bare (50%), weeds (50%).	05051D	.82	4.38	177	5	4.69	2.11	70
			05051W	2.84	6.01	145	30	3.25	3.15	
Do	B	......do......	05052D	1.02	3.64	120	4	6.25	3.16	65
			05052W	2.71	5.94	105	7	5.17	1.31	
Fuquay pebbly loamy sand.	B	Grass (80%), bare (20%).	06061D	1.76	5.22	164	3	4.69	3.77	60
			06061W	2.43	5.83	92	3	4.45	2.82	
Do	B	......do......	06062D	1.63	4.96	163	12	2.64	2.31	61
			06062W	2.64	5.92	89	4	4.33	3.17	
Do	B	......do......	06064D	1.90	5.64	120	9	3.13	2.36	30
			06064W	2.71	5.78	305	2	6.25	2.65	
Kershaw coarse sand.	A	Bare (60%), weeds (40%).	07071D	.45	1.25	130	7	6.13	6.11	35
			07071W	2.47	4.60	105	12	6.13	6.08	
Do	A	......do......	07072D	.45	1.48	265	4	6.25	6.24	...
Leefield loamy sand.	C	Weeds (80%), bare (20%).	08081D	1.08	3.95	120	2	6.50	1.92	100
			08081W	2.87	5.30	100	5	5.17	.96	
Do	C	......do......	08082D	1.30	4.34	150	9	4.81	2.81	65
			08082W	2.77	5.52	140	14	3.00	2.38	
Robertsdale loamy sand.	C	Weeds (70%), bare (30%).	09091D	1.79	5.51	120	4	4.81	.95	60
			09091W	2.42	6.26	120	4	3.25	.89	
Do	C	......do......	09092D	2.00	5.67	120	2	4.81	1.12	65
			09092W	2.65	6.32	150	2	6.73	1.29	
Stilson loamy sand.	B	Weeds (50%), bare (50%).	10101D	1.22	5.18	117	4	6.25	2.71	60
			10101W	2.80	6.38	91	8	5.05	1.17	
Do	B	......do......	10102D	1.19	4.48	180	6	4.93	3.17	75
			10102W	2.96	6.04	109	11	3.00	1.91	
Tifton loamy sand.	B	Weeds (90%), bare (10%).	13122D	1.58	6.14	120	7	6.13	3.58	60
			13122W	2.60	6.90	125	5	4.57	2.51	
Do	B	......do......	14121D	1.94	6.38	120	6	4.69	1.82	60
			14121W	2.97	7.22	120	4	6.37	.2	
Do	B	......do......	14122W	3.06	7.35	120	8	2.74	.67	...
Do	B	......do......	15121D	1.75	6.52	120	4	4.69	2.29	60
			15121W	3.29	7.29	120	4	6.37	.94	
Do	B	......do......	15122D	1.54	7.54	120	4	4.81	3.62	60
			15122W	3.51	9.31	120	5	2.64	1.83	

See footnotes at end of table.

TABLE 2.—*Infiltration data*—Continued

Soil series	Hydrologic group[1]	Cover conditions	Identification code[2]	Soil moisture (in) at— 0"–12"	12"–36"	Total length applied rainfall (min)	Time from start of applied rainfall to start of runoff (min)	Rainfall intensity (in/h)	Final infiltration rate (in/h)	Time between applied rainfall (min)
Troup sand	A	Grass (100%)	11111D	1.13	5.27	.34	4	4.57	1.91	63
			11111W	2.73	6.44	112	5	6.50	1.12	
Do	A	do	11112D	1.13	4.91	150	5	2.64	2.41	60
			11112W	2.36	6.10	123	5	4.57	1.73	
Do	A	do	11113D	1.03	3.88	120	3	5.17	4.20	63
			11113W	2.31	5.63	120	17	2.76	2.33	
Do	A	Weeds (60%), bare (40%)	12112D	1.73	3.39	150	3	6.25	2.78	78
			12112W	2.10	5.69	90	6	6.25	1.63	
Do	A	do	12113D	2.03	5.59	150	6	6.50	2.18	76
			12113W	2.87	6.73	88	4	2.88	1.27	
Do	A	do	12114D	2.65	8.57	204	4	3.85	1.66	76
			12114W	3.38	9.80	60	2	3.85	1.01	

[1] Soil Conservation Service hydrologic classification (*24*). [2] D=plot conditions before water application, W=plot conditions after water application, and WW=plot conditions at time of 3d infiltration run. See appendix for explanation of identification code.

runoff rate was obtained. For each application, 1 to 2½ h of rainfall were required to produce constant runoff.

A Troxler model 200–B scaler and model 104–A depth probe with a 100-mc, 241 Am–Be neutron source were used to measure soil moisture. Neutron soil-moisture readings were made at 6-in (15.24-cm) intervals to a total soil depth of 36 to 48 in (91.4 to 121.9 cm) at the beginning and end of each rainfall event. The depth to which the neutron probe was read depended on the location of the clay layer (*16*).

Soil samples for gravimetric moisture determination were taken from two holes simultaneously with the initial neutron measurements. The holes were located within a radius of 2 to 4 ft (61.00 to 122.00 cm) of the neutron probe access tube, and an orchard auger was used to take the fragmented samples. The radius for gravimetric sampling was chosen to minimize site disturbance.

The soil samples were taken every 3 in (7.60 cm) to a total depth of 42 in (106.70 cm), placed in soil-moisture cans, and brought into the laboratory and weighed. They were then dried in a forced-draft oven at 221° F (105° C) for at least 24 h and weighed again. Tests indicated that drying beyond 24 h produced negligible additional water loss. Soil moisture was initially calculated as percent of dry weight, then converted into percent by volume by use of bulk-density data previously obtained for each soil horizon at each site (*10*).

A single volumetric soil-water value was determined for each neutron-value reading by averaging the soil-water volumetric data for the thickness measured by the neutron soil-water meter. These data were used to field-calibrate the neutron probe. Rawls and Asmussen (*15*) give the results of the field calibration.

DETERMINATION OF INFILTRATION RATES

Infiltration data are summarized in table 2. Final infiltration rates were determined by subtracting the ultimate surface runoff rate from the rainfall rate. Soil-profile descriptions, moisture-tension data, and detailed infiltration data for each site are given in the appendix.

A continuity equation was used to obtain infiltration values from the rainfall-runoff data.

For this study, the continuity equation becomes

$$I = R - RO - S,\quad (8)$$

where I =volume of infiltration (inches),
R =volume of rainfall (inches),
RO =volume of runoff (inches),
and S =surface storage (assumed negligible except at the beginning of the study),

When time is included in equation 8, infiltration rates can be determined. Evaporation was considered to be negligible in this study and was therefore ignored.

The continuous interpolation method described by Snyder (19) was used to reduce the rainfall-runoff values to instantaneous infiltration rates. This method was also modified to compute beginning and ending infiltration rates.

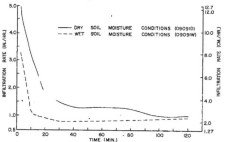

FIGURE 2.—Infiltration-capacity curves for Robertsdale loamy sand.

ANALYSIS

The equations tested in this study (equations 2, 3, 5, 6, and 7) are of the exponential form. However, none of the infiltration runs on Kershaw coarse sand produced results expressible in exponential form. Because the infiltration rates of this coarse sand exceeded the capability of the infiltrometer to apply rainfall, all Kershaw run data were excluded from analysis. Additionally, a few runs on other soils did not produce exponential results, and therefore these data were excluded from analysis. All infiltration runs used in the analysis are listed in table 3. Typical infiltration-capacity curves for dry and wet initial soil-moisture conditions are shown in figure 2. As can be seen in the figure, the curves generally reflect a high initial infiltration rate followed by a rapid decline to a comparatively constant rate.

TABLE 3.—*Infiltration runs[1] used for analysis*

Soil type and identification code[2]	Soil type and identification code[2]
Alapaha loamy sand:	Leefield loamy sand:
01011D	08081W
01011W	08082D
01012D	08082W
01012W	Robertsdale loamy sand:
Carnegie loam sandy:	09091D
02021W	09091W
02022W	09092D
Cowarts loamy sand:	09092W
03031WW	Stilson loamy sand:
03032W	10101W
03032WW	10102W
Cowarts (Z) loamy sand:	Tifton loamy sand:
16031W	13122D
16032W	13122W
16033W	14121D
17031W	14121W
17032W	14122W
17033WW	15121D
Dothan loamy sand:	15121W
04041W	15122D
Fuquay loamy sand:	Troup sand:
05051W	11111D
05052W	11112W
Fuquay pebbly	11113W
loamy sand:	12112D
06062D	12112W
06062W	12113D
06064D	12113W
06064W	12114D
	12114W

[1] Total of 48 runs.
[2] See appendix for explanation of the identification code.

The pattern-search method of optimization (5) was used to fit the infiltration equations to the data. The optimization criterion used was minimization of the sum of squares of error between the observed and predicted values.

During the early stages of rainfall, the difference between rainfall and runoff does not represent the true amount of water infiltrating into the soil profile. A portion of early rainfall is required to satisfy surface storage and interception. The first two infiltration values from each test series were therefore omitted in the fitting process.

Horton's three-parameter infiltration equation (equation 2) fits the data well, with an arithmetic mean correlation coefficient of 0.95. The results of the fittings are summarized by

(Continued on page 14.)

9

TABLE 4.—*Summary of fitting Horton's equation*[1]

Parameter	Alapaha (01) N=4	Carnegie (02) N=2	Cowarts (03) N=3	Cowarts Z (03,16,17) N=6	Dothan (04) N=1	Fuquay (05) N=2	Fuquay (06) N=4	Leefield (08) N=3	Robertsdale (09) N=4	Stilson (10) N=2	Tifton (13,14,15) N=8	Troup (11,12) N=9	Average
f_c													
Average	1.04	1.77	1.95	2.23	2.63	2.42	2.59	1.73	1.18	1.55	1.63	1.80	1.88
Minimum	.37	1.60	1.66	.98	....	1.71	2.29	1.11	.86	1.19	.33	1.08	.33
Maximum	2.40	1.93	2.11	2.66	....	3.14	3.33	2.14	1.37	1.91	3.51	2.75	3.51
f_o													
Average	19.00	14.77	15.28	17.25	3.47	6.24	38.76	11.34	12.41	8.11	9.67	23.01	14.94
Minimum	5.90	8.73	5.27	3.28	....	3.74	21.98	3.15	6.55	4.06	4.37	3.43	3.15
Maximum	23.87	20.81	33.70	66.77	....	8.74	75.50	26.52	20.98	12.17	31.13	116.68	116.68
K_f													
Average	38.29	19.64	10.65	16.12	1.40	4.70	39.98	7.70	21.75	5.66	7.28	32.71	7. ⬤
Minimum	6.40	10.65	2.09	5.04	....	4.07	22.35	.37	8.37	3.66	.47	2.40	.37
Maximum	91.04	28.62	24.42	57.86	....	5.33	57.65	2.08	32.22	7.67	6. ⬤	100.11	100.11
Correlation coefficient:													
Mean	.98	.92	.99	.96	.98	.86	.87	.96	.97	.99	.96	.97	.95
Minimum	.94	.89	.98	.91	....	.75	.80	.90	.94	.99	.92	.86	.75
Maximum	.99	.94	.99	.98	....	.97	.94	.99	.99	.99	.99	.99	.99

[1] See soils location code, appendix, for explanation of number following soil name. N = number of samples. Horton's equation is equation 2 in text.

TABLE 5.—*Summary of fitting Green and Ampt's equation*[1]

Parameter	Soil series												
	Alapaha (01) N=4	Carnegie (02) N=2	Cowarts (03) N=3	Cowarts Z (03,16,17) N=6	Dothan (04) N=1	Fuquay (05) N=2	Fuquay (06) N=4	Leefield (08) N=3	Robertsdale (09) N=4	Stilson (10) N=2	Tifton (13,14,15) N=8	Troup (11,12) N=9	Average
A													
Average	0.37	1.37	2.09	1.73	2.71	0.96	2.51	1.15	0.67	0.60	1.09	1.31	1.38
Minimum	−.28	1.29	.94	−.08	....	.60	2.24	−.05	.44	−.49	−2.51	.17	−2.51
Maximum	1.76	1.45	3.30	3.59	....	1.32	3.24	2.38	.98	1.68	3.76	2.52	3.76
C													
Average	.99	1.02	.81	1.54	.30	4.63	.24	2.51	.82	2.21	2.27	1.41	1.56
Minimum	.42	.66	.15	.09	....	3.05	.10	.70	−.45	.78	.70	.36	.09
Maximum	1.67	1.40	1.92	3.76	....	6.20	.50	5.10	1.20	3.64	6.90	5.20	6.90
Correlation coefficient:													
Mean	.84	.75	.93	.90	.84	.77	.61	.72	.93	.97	.76	.81	.81
Minimum	.61	.55	.84	.63	....	.61	.44	.54	.85	.96	.51	.62	.44
Maximum	.98	.94	.98	.99	....	.94	.76	.85	.98	.99	.95	.97	.99

[1] See soils location code, appendix, for explanation of number following soil name. *N*=number of samples. Green and Ampt's equation is equation 4.

TABLE 6.—*Summary of fitting Phillip's equation*[1]

Parameter	Soil series												
	Alapaha (01) N=4	Carnegie (02) N=2	Cowarts (03) N=3	Cowarts Z (03,16,17) N=6	Dothan (04) N=1	Fuquay (05) N=2	Fuquay (06) N=4	Leefield (08) N=3	Robertsdale (09) N=4	Stilson (10) N=2	Tifton (13,14,15) N=8	Troup (11,12) N=9	Average
S													
Average	0.93	0.82	0.84	0.85	0.33	1.08	0.23	0.74	0.79	1.30	1.01	0.88	0.82
Minimum	.58	.75	.51	.11	....	.30	.05	.33	.55	.70	.29	.34	.05
Maximum	1.54	.89	1.36	1.86	....	1.86	.49	1.13	1.00	1.90	1.52	2.24	2.24
C													
Average	.08	.98	1.38	1.47	2.51	1.34	2.39	1.60	.36	.29	1.19	.99	1.21
Minimum	−.96	.86	.29	−.42	....	−.21	2.08	−.14	.21	−.74	−1.32	−.01	−1.32
Maximum	1.43	1.10	2.64	3.42	....	2.90	3.03	2.64	.62	1.31	2.46	2.30	3.42
Correlation coefficient:													
Mean	.86	.75	.91	.88	.93	.69	.57	.85	.92	.96	.89	.83	.82
Minimum	.71	.59	.85	.62	....	.42	.37	.83	.85	.94	.74	.70	.37
Maximum	.93	.91	.97	.98	....	.96	.75	.89	.97	.98	.97	.99	.99

[1] See soils location code, appendix, for explanation of number following soil name. *N*=number of samples. Phillip's equation is equation 5 in text.

TABLE 7.—Summary of fitting Holtan's equation[1]

Parameter	Alapaha (01) N=4	Carnegie (02) N=2	Cowarts (03) N=3	Cowarts Z (03,16,17) N=6	Dothan (04) N=1	Fuquay (05) N=2	Fuquay (06) N=4	Leefield (08) N=3	Robertsdale (09) N=4	Stilson (10) N=2	Tifton (13,14,15) N=8	Troup (11,12) N=9	Average
α													
Average	0.11	0.14	0.24	0.18	0.09	0.06	0.08	0.16	0.12	0.24	0.17	0.15	0.14
Minimum	.04	.11	.20	.05	....	.05	-.01	.05	.01	.19	.04	.08	-.01
Maximum	.15	.18	.27	.29	....	.07	.18	.32	.19	.28	.38	.28	.38
S_t													
Average	5.70	9.92	8.37	10.42	15.50	18.06	18.13	10.91	7.03	7.50	8.28	9.92	10.60
Minimum	2.95	7.96	8.12	4.58	....	7.56	12.75	4.37	6.70	5.50	2.20	2.18	2.18
Maximum	9.61	11.88	8.63	19.50	....	28.56	22.50	17.00	7.31	9.50	20. 82	18.5	28.56
n													
Average	2.45	1.15	1.43	1.31	1.09	1.38	.84	1.49	1.88	1.47	2.24	1.66	1.53
Minimum	1.29	1.12	1.29	.66	....	.64	.35	.84	1.47	1.12	.72	.82	.35
Maximum	4.48	1.17	1.56	2.92	....	2.13	1.11	2.66	2.51	1.82	7.27	5.25	7.27
f_c													
Average	0.45	0.53	-0.97	1.05	1.59	1.18	1.90	1.01	-0.56	-0.40	0.76	0.47	0.59
Minimum	-.04	.40	-1.75	-.53	....	-.40	1.45	.44	-1.18	-1.21	-.01	-1.40	-1.75
Maximum	.89	.67	-.58	2.25	....	2.77	2.39	1.75	-.10	.41	1.92	2.15	2.77
Correlation coefficient:													
Mean	.59	.47	.86	.78	.92	.88	.68	.83	.64	.83	.83	.62	.74
Minimum	.32	.27	.78	.67	....	.87	.60	.61	.49	.82	.28	.30	.27
Maximum	.64	.67	.98	.90	....	.89	.70	.98	.80	.84	.98	.90	.98

[1] See soils location code, appendix, for explanation of number following soil name. N=number of samples. Holtan's equation is equation 6 in text.

TABLE 8.—Summary of fitting Snyder's equation[1]

Parameter		Alapaha (01) N 4	Carnegie (02) N 2	Cowarts (03) N 3	Cowarts Z (03,16,17) N 6	Dothan (04) N 1	Fuquay (05) N 2	Fuquay (06) N 4	Soil series Leefield (08) N 3	Robertsdale (09) N 4	Stilson (10) N 2	Tifton (13,14,15) N 8	Troup (11,12) N 9	Average
f_o	Average	2.66	2.58	3.32	3.28	3.23	3.03	2.71	3.27	2.80	2.60	3.29	2.77	2.96
	Minimum	1.66	2.28	2.49	2.10	....	2.97	2.32	2.75	1.90	2.40	1.85	1.68	1.66
	Maximum	4.06	2.88	4.26	4.96	....	3.13	3.60	4.15	3.19	2.80	4.29	3.52	4.96
a	Average	2.76	.48	.14	4.35	.35	1.35	1.06	.86	4.20	.01	1.54	.82	1.49
	Minimum	.01	.19	.01	.01	....	1.00	.01	.22	.01	.01	.01	.01	.01
	Maximum	11.03	.77	.35	14.20	....	1.70	1.50	1.33	7.69	.01	2.44	3.24	14.20
b	Average	−14.96	−5.44	−3.89	−6.70	−.56	−.40	−1.51	−3.89	−10.55	−1.77	−7.45	−9.26	−5.51
	Minimum	−20.60	−6.96	−6.68	−19.35	....	−.50	−3.17	−10.38	−14.92	−2.54	−24.20	−22.08	−24.20
	Maximum	−6.02	−3.33	−1.61	−.59	....	−.30	−.60	−.55	−7.96	−1.00	−.33	−2.21	−.30
f_a	Average	8.00	8.00	7.59	7.78	7.80	7.80	7.83	7.60	8.00	8.00	7.68	7.98	7.84
	Minimum	7.99	7.99	6.79	7.27	....	7.60	7.30	6.80	7.99	7.99	5.75	7.85	5.75
	Maximum	8.00	8.00	8.00	8.00	....	8.00	8.00	8.00	8.00	8.00	8.00	8.00	8.00
f_r	Average	1.00	1.75	1.52	1.56	2.53	.22	.83	1.10	1.09	1.90	.94	1.37	.62
	Minimum	.42	1.56	.83	.54	....	.10	.53	.85	.84	1.60	.39	.53	.10
	Maximum	2.26	1.93	2.12	2.91	....	.34	1.00	1.37	1.26	2.20	2.23	1.87	2.91
Correlation coefficient:	Mean	.97	.91	.99	.87	.96	.42	.51	.96	.98	.99	.87	.92	.86
	Minimum	.94	.85	.98	.72	....	.40	.20	.90	.97	.98	.48	.63	.20
	Maximum	.99	.97	.99	.98	....	.44	.78	.99	.98	.99	.99	.99	.99

[1] See soils location code, appendix, for explanation of number following soil name. N=number of samples. Snyder's equation is equation 7 in text.

soil type in table 4. As shown in table 4, there is wide variation in equation parameters, which results from factors other than differences in soil types alone. The K_f values were the most variable because, in general, the steepness of the initial portion of the infiltration curve controls the K_f value, which is dependent on initial soil-moisture conditions. The f_a values indicate a very high initial infiltration rate, which is realistic for the sandy soils. The parameter f_c, with a few exceptions, agreed closely with the final infiltration rate. Final infiltration rate could be calculated with an average error of 0.13 in/h (3.30 mm/h).

The results of fitting Green and Ampt's (equation 4) and Phillip's (equation 5) equations to the data are summarized in tables 5 and 6, respectively. The parameters for these equations are also highly variable. Both equations consistently overestimate the early portion of the infiltration curve and underestimate the later portion. Therefore, the final infiltration rate was consistently underestimated, with an average error of 0.18 in/h (4.57 mm/h). Furthermore, the fit precision was highly variable, indicating that the equations could not consistently fit all the data.

The results of fitting Holtan's equation (equation 6) to the data are summarized in table 7. The range of a values was reasonable for the vegetation encountered. However, there was no general agreement between the fitted a values and the observed surface conditions of the infiltrometer plots. Apparently, other surface variables are needed to provide a good estimate of the a value. A clay layer caused much of the infiltrating water to become lateral flow, and resulted in very high S_t values which were unrealistic for the soils studied (16). The n values were close to the 1.4 used by Holtan (7). Some f_c values were negative, which is unrealistic, because this value is designated as the steady state infiltration rate. Since Holtan specified that n equals 1.4 and f_c equals a small rate (0 to 0.30 in/h), based on the Soil Conservation Service's hydrologic soil classification (23, 24), a second fitting was made, in which n equaled 1.4 and f_r equaled values specified according to hydrologic grouping. The two-parameter equation fits the data about as well as the four-parameter equation. Both equations predicted the final infiltration rate with an average error of 0.30 in/h. The two-parameter equation consistently overesti-

mated the early portion of the infiltration curve and underestimated the later portion.

Preliminary analysis of Snyder's equation (equation 7) indicated that certain restrictions had to be placed on some of the parameters to obtain good fits. The following are the restrictions: (1) $a \geq 0.0$, (2) $b \leq 0.0$, and (3) $0.0 < f_a \leq 8.0$. The results of the fittings are summarized in table 8. The initial infiltration rate, f_{t-1} (at $t = 1$), was fairly stable. Varying initial soil-moisture conditions resulted in variability of initial infiltration rates in individual soil series. The variability of the shape coefficients was extreme, and not explainable by physical conditions. The f_a value generally went to the upper limit of 8.0 in/h (20.32 cm/h). The f_r value tended to agree with the final infiltration rates for the runs. Except for the Fuquay soil series, Snyder's equation fit the data well, with an arithmetic mean correlation coefficient of 0.95. Moreover, the final infiltration rate was predicted with an average error of 0.11 in/h.

LITERATURE CITED

(1) Amerman, C. R. 1969. Finite difference solutions of unsteady two-dimensional partially saturated porous media flow. 136 pp. Ph. D. dissertation, Purdue University, Lafayette.

(2) Bertrand, A. R., and Parr, J. F. 1961. Design and operation of the Purdue sprinkling infiltrometer. Purdue Agric. Exp. Stn. Res. Bull. 723, 30 pp.

(3) Brasher, B. R., Franzmeier, D. P., Valassis, V., and Davidson, S. E. 1966. Use of saran resin to coat natural soil clods for bulk-density and water-retention measurement. Soil Sci. 101: 108.

(4) Dixon, R. M., and Peterson, A. E. 1964. Construction and operation of a modified spray infiltrometer and flood infiltrometer. Wis. Agric. Exp. Stn. Res. Rep. 15, 31 pp.

(5) Green, R. R. 1970. Optimization by the pattern search method. Tenn. Val. Auth. Res. Pap. No. 7, 73 pp.

(6) Green, W. H., and Ampt, G. 1911. Studies of soil physics. Part I. The flow of air and water through soils. J. Agric. Sci. 4: 1–24.

(7) Holtan, H. N. 1961. A concept for infiltration estimates in watershed engineering. U.S. Dep. Agric., Agric. Res. Serv. [Rep.] ARS 41–51, 25 pp.

(8) ———, England, C. B., and Shanholtz, V. O. 1967. Concepts in hydrologic soil grouping. Trans. ASAE (Am. Soc. Agric. Eng.) 10: 407–410.

(9) Horton, R. E. 1940. An approach toward physical interpretation of infiltration capacity. Soil Sci. Soc. Am. Proc. 5: 399–417.

(10) McCreery, R. A. 1967. Notes on data for soils

from Little River watershed, Tift and Turner Counties, Georgia. Rep. to Southeast Watershed Research Center, Athens, Ga. 47 pp. University of Georgia, Athens.

Overton, D. E. 1964. Mathematical refinement of an infiltration equation for watershed engineering. U.S. Dep. Agric., Agric. Res. Serv. [Rep.] ARS 41–99, 11 pp.

Phillip, J. R. 1954. An infiltration equation with physical significance. Soil Sci. 77: 153–157.

———. 1957. Numerical solution of equations of the diffusion-type with diffusivity concentration-dependent, II. Aust. J. Phys. 10: 29–42.

———. 1957. The theory of infiltration. 1. The infiltration equation and its solution. Soil Sci. 83: 345–357.

Rawls, W. J., and Asmussen, L. E. 1974. Neutron probe field calibration for soils in the Georgia Coastal Plain. Soc. Sci. 116(4): 262–265.

———. 1973. Subsurface flow in Georgia Coastal Plain. J. Irrig. Drain. Div., Proc. Am. Soc. Civ. Eng. 99(IR3): 375–386.

Skaggs, R. W., Huggins, L. F., Monke, E. J., and Foster, G. R. 1969. Experimental evaluation of infiltration equations. Trans. ASAE (Am. Soc. Agric. Eng.) 12(6): 822–828.

(18) Smith, R. E. 1971. Discussion: "A proposed watershed retention function" by W. M. Snyder. J. Irrig. Drain Div., Proc. Am. Soc. Civ. Eng. 97(IR3): 544–545.

(19) Snyder, W. M. 1967. Extended continuous interpolation. J. Hydraul. Div., Proc. Am. Soc. Civ. Eng. 93(HY5): 261–280.

(20) ———. 1971. A proposed watershed retention function. J. Irrig. Drain. Div., Proc. Am. Soc. Civ. Eng. 97(IR1): 193–201.

(21) U.S. Department of Agriculture. 1954. Diagnosis and improvement of saline and alkali soils. Agric. Handb. No. 60, 160 pp.

(22) ———. 1962. Soil survey manual. Agric. Handb. No. 18, 503 pp.

(23) U.S. Department of Agriculture, Soil Conservation Service. 1959. Soil survey, Tift County, Ga. Series 1946, No. 3, 28 pp.

(24) ———. 1964. National engineering handbook. Section 4, Hydrology, Chapter 7, 7.1–7.5.

(25) ———. 1967. Little River watershed — Tift and Turner Counties, Ga. — 44 soil-profile descriptions — 2,999 soil samples, 83 pp.

15

APPENDIX.—SOIL-PROFILE DESCRIPTIONS, MOISTURE-TENSION DATA, AND INFILTRATION-RUN DATA

This appendix supplies a soil-profile description, a table summarizing moisture-tension data, and a printout of infiltration-run data for each of the 17 sites involved in the present study. The sequence, as it appears in the previous sentence, is repeated as each distinct location is introduced, with the exception of the soil-profile description and moisture-tension table for Cowarts loamy sand, which are identical for locations 03, 16, and 17.

Soil-profile descriptions are compiled from those prepared by John W. Calhoun, soil scientist, Soil Conservation Service, U.S. Department of Agriculture (USDA), in accordance with USDA Handbook No. 18, "Soil Survey Manual" (22).

Moisture-holding capacity was measured, using pressure membranes and pressure plates as described by methods 29, 30, 31, and 32 in USDA Handbook No. 60 (21), with modifications. Bulk densities were determined by using the saran coating method (3). A computer program previously developed to reduce data for the U.S. Hydrograph Laboratory was used for data calculations.

Moisture-tension data are explained in the following paragraphs. Depths shown refer to total depth from the soil surface to the top of the subject horizon.

Three lines of data are given for each profile. The first five columns on line 1 list the equilibrium moisture content by weight at 0.1, 0.3, 0.6, 3.0, and 15.0 bars tension, respectively. The same columns of line 2 list the equilibrium moisture contents in percent by volume for the same tensions. The sixth column gives the bulk density of the soil in the form indicated by the footnotes to each profile. The bulk density figures shown on lines 1 and 2 are for values calculated at 0.3 bar tension and for ovendry conditions, respectively.

Pore space values are given in column 7. Values were obtained by using standard calculation techniques, using the observed bulk density results and assuming a particle density of 2.65 grams per cubic centimeter.

On the third line, the FRAGMENT datum is the moisture content at 0.3 bar tension, in percent by weight, of the large sample used for bulk density determination. SIEVED is the moisture content at 15.0 bars tension,

in percent by weight, of a subsample remaining after al material larger than 2 mm in diameter has been screenec out. ROCK PERCENT is the percentage of materia larger than 2 mm in diameter which was removed by sieving, expressed on a whole sample (as received) basis

Footnotes to the bulk density values indicate the type of sample used for the bulk density determination. FIST denotes undisturbed saran-coated fragments; CORE means no core samples were taken; LOOSE indicate: that the particular horizon was not sufficiently cohesiv to obtain satisfactory fragments for coating and, there fore, bulk density was determined by consolidation of saturated sample in the laboratory.

The identification code used in this appendix is th same as that used in the text. Each number consists o five units; and a code number, such as 01011D, is broke down as follows:

Location	Soil type	Plot	Soil-moisture condition
01	01	1	D

Location: See figure 2 for location of each site.

Soil type:

01	Alapaha loamy sand
02	Carnegie sandy loam
03	Cowarts loamy sand
04	Dothan loamy sand
05	Fuquay loamy sand
06	Fuquay pebbly loamy sand
07	Kershaw coarse sand
08	Leefield loamy sand
09	Robertsdale loamy sand
10	Stilson loamy sand
11	Troup sand
12	Tifton loamy sand

Plot: 1 = first plot, 2 = second plot, 3 = third plot, an 4 = fourth plot. Order is arbitrary and for identificatio purposes only.

Soil-moisture condition: D = first infiltration run o the plot, W = second infiltration run on the plot, an WW = third infiltration run on the plot.

Soil-profile descriptions, moisture-tension data, an infiltration-run data appear below.

ALAPAHA LOAMY SAND (01)

Location: 1 mi north of livestock barn on Coastal Plain Experiment Station along field road; west along field road for 0.3 mi; 200 ft north of road in idle area; Tift County, Ga.

Land use or cover: Idle—myrtle, wiregrass, gallberry.

Topography: Nearly level — less than 1% slope.

Great soil group: Arenic plinthic palequults; loamy, siliceous, thermic.

Parent material: Unconsolidated marine sediment of sandy clay loam.

Drainage: Poorly drained.

Horizon and Description

A1: 0 to 7 inches. Dark-gray (N/4) loamy sand with few fine faint mottles of light gray; weak, fine granular structure; very friable, nonsticky; many fine roots; very strongly acid; abrupt smooth boundary.

A2: 7 to 32 inches. Gray (10YR–5/1) loamy sand with a few fine faint mottles of light gray; weak, fine granular structure; very friable, nonsticky; fine and medium roots common; some clean sand grains; very strongly acid; clear, wavy boundary.

B21tg: 32 to 38 inches. Light-gray (10YR–7/1) sandy clay loam with few fine faint mottles of light yellowish brown and yellowish red; weak, medium subangular blocky structure; friable, slightly sticky; very strongly acid; gradual wavy boundary.

B22tp1: 38 to 48 inches. Yellowish brown (10YR–5/8) sandy clay loam with many coarse, distinct, and prominent mottles of light-gray (10YR–7/1) and yellowish red (5YR–5/8); weak, medium subangular blocky structure; matrix firm in place, crushes to friable mass; soft plinthite 10% to 20% by volume; very strongly acid; gradual wavy boundary.

B23tp1: 48 to 65 inches. Brownish-yellow (10YR–6/6) sandy clay loam with many coarse, distinct, and prominent mottles of light gray (10YR–7/1), red (10YR–4/8), and strong brown (7.5YR–5/8); red mottles increase with depth; moderate, medium subangular blocky structure; matrix firm in place, crushes to friable mass; soft plinthite 10% to 30% by volume; very strongly acid.

Remarks: Colors are given for moist soil. Reaction determined by Soiltex.

WEIGHT PERCENT AND VOLUME PERCENT OF WATER RETAINED

DEPTH (inches)	TENSIONS (BARS)					BD G/CC	TP PCT	K
	.1	.3	.6	3.	15.			
0–7	8.54	6.29	4.49	3.36	3.14	1.49[1]	43.77	2.00–6.30
	12.72	9.37	6.69	5.01	4.68	1.49	43.77	
	FRAGMENT	5.64		SIEVED	2.27	ROCK PCT	1.68	
7–32	5.72	3.34	2.77	1.62	0.68	1.61[1]	39.25	2.00–6.30
	9.21	5.38	4.46	2.61	1.09	1.60	39.62	
	FRAGMENT	2.03		SIEVED	0.77	ROCK PCT	0.90	
32–38	12.14	8.16	6.92	5.44	3.94	1.47[1]	44.53	0.06–0.20
	17.85	12.00	10.17	8.00	5.79	1.61	39.25	
	FRAGMENT	7.07		SIEVED	4.62	ROCK PCT	4.40	
38–48	11.42	9.71	6.04	5.91	4.81	1.77[1]	33.21	0.06–0.20
	20.21	17.19	10.69	10.46	8.51	1.79	32.45	
	FRAGMENT	7.10		SIEVED	4.13	ROCK PCT	3.56	
48+	13.78	10.09	9.33	8.78	5.33	1.73[1]	34.72	0.06–0.20
	23.84	17.46	16.14	15.19	9.22	1.76	33.58	
	FRAGMENT	8.28		SIEVED	5.16	ROCK PCT	4.56	

1=FIST
2=CORE
3=LOOSE

SOIL TYPE - ALAPAHA LOAMY SAND
IDENTIFICATION CODE - 01011D
COVER - WEEDS-90, BARE-10
DATE OF RUN - 10 31 69
RAINFALL INTENSITY - 6.730 INCHES/HOUR
INITIAL SOIL MOISTURE FOR THE 0 TO 12 INCH DEPTH - 2.30 INCHES
INITIAL SOIL MOISTURE FOR THE 12 TO 36 INCH DEPTH - 5.18 INCHES
FINAL SOIL MOISTURE FOR THE 0 TO 12 INCH DEPTH - 3.10 INCHES
FINAL SOIL MOISTURE FOR THE 12 TO 36 INCH DEPTH - 6.01 INCHES

TIME FROM START OF RAIN (MINUTES)	ACCUMULATED RUNOFF (INCHES)	RUNOFF RATE (IN/HR)	ACCUMULATED INFILTRATION (INCHES)	INFILTRATIO RATE (IN/HR)
3	0.000	0.000	0.336	6.730
5	0.072	3.124	0.440	3.605
10	0.425	3.292	0.696	3.438
15	0.721	3.628	0.960	3.101
20	1.035	3.953	1.207	2.776
25	1.373	4.130	1.430	2.600
30	1.720	4.163	1.645	2.567
35	2.064	4.164	1.862	2.565
40	2.415	4.242	2.071	2.487
45	2.765	4.236	2.282	2.494
50	3.112	4.219	2.496	2.510
55	3.479	4.346	2.690	2.384
60	3.851	4.455	2.879	2.275
65	4.213	4.427	3.077	2.303
70	4.585	4.393	3.267	2.336
75	4.952	4.407	3.460	2.323
80	5.323	4.490	3.650	2.240
85	5.682	4.349	3.852	2.381
90	6.055	4.353	4.040	2.377
95	6.443	4.521	4.212	2.208
100	6.809	4.448	4.408	2.282
105	7.185	4.465	4.593	2.264
110	7.552	4.414	4.787	2.316
115	7.932	4.534	4.967	2.195
120	8.307	4.562	5.153	2.168
125	8.669	4.404	5.352	2.325
130	9.060	4.645	5.521	2.085
135	9.443	4.761	5.700	1.969
140	9.804	4.627	5.899	2.102

SOIL TYPE — ALAPAHA LOAMY SAND
IDENTIFICATION CODE — 01011W
COVER — WEEDS-90, BARE-10
DATE OF RUN — 10 31 69
RAINFALL INTENSITY — 4.447 INCHES/HOUR
INITIAL SOIL MOISTURE FOR THE 0 TO 12 INCH DEPTH — 3.36 INCHES
INITIAL SOIL MOISTURE FOR THE 12 TO 36 INCH DEPTH — 5.89 INCHES
FINAL SOIL MOISTURE FOR THE 0 TO 12 INCH DEPTH — 3.46 INCHES
FINAL SOIL MOISTURE FOR THE 12 TO 36 INCH DEPTH — 5.79 INCHES

TIME FROM START OF RAIN (MINUTES)	ACCUMULATED RUNOFF (INCHES)	RUNOFF RATE (IN/HR)	ACCUMULATED INFILTRATION (INCHES)	INFILTRATION RATE (IN/HR)
3	0.000	0.000	0.222	4.447
5	0.052	3.485	0.262	0.961
10	0.408	3.602	0.332	0.844
15	0.697	3.613	0.414	0.833
20	1.003	3.704	0.479	0.742
25	1.310	3.702	0.542	0.744
30	1.618	3.701	0.605	0.745
35	1.928	3.730	0.666	0.716
40	2.239	3.738	0.725	0.708
45	2.543	3.684	0.791	0.762
50	2.850	3.697	0.855	0.749
55	3.156	3.631	0.920	0.815
60	3.466	3.602	0.981	0.844
65	3.767	3.559	1.049	0.887
70	4.064	3.564	1.123	0.882
75	4.363	3.538	1.195	0.908
80	4.663	3.587	1.265	0.859
85	4.968	3.624	1.331	0.822
90	5.266	3.625	1.404	0.821
95	5.564	3.565	1.477	0.881
100	5.853	3.443	1.558	1.003
105	6.165	3.541	1.616	0.905
110	6.481	3.742	1.671	0.704
115	6.763	3.540	1.760	0.906
120	7.082	3.730	1.812	0.716

```
SOIL TYPE - ALAPAHA LOAMY SAND
IDENTIFICATION CODE - 01012D
COVER - WEEDS-90, BARE-10
DATE OF RUN - 11 01 69
RAINFALL INTENSITY - 4.567 INCHES/HOUR
INITIAL SOIL MOISTURE FOR THE 0 TO 12 INCH DEPTH - 2.33 INCHES
INITIAL SOIL MOISTURE FOR THE 12 TO 36 INCH DEPTH - 6.64 INCHES
FINAL SOIL MOISTURE FOR THE 0 TO 12 INCH DEPTH - 3.64 INCHES
FINAL SOIL MOISTURE FOR THE 12 TO 36 INCH DEPTH - 6.38 INCHES
```

TIME FROM START OF RAIN (MINUTES)	ACCUMULATED RUNOFF (INCHES)	RUNOFF RATE (IN/HR)	ACCUMULATED INFILTRATION (INCHES)	INFILTRATION RATE (IN/HR)
4	0.000	0.000	0.304	4.567
5	0.020	1.442	0.360	3.124
10	0.180	2.473	0.580	2.093
15	0.380	2.630	0.761	1.937
20	0.621	3.128	0.901	1.438
25	0.902	3.658	1.000	0.908
30	1.222	3.898	1.060	0.669
35	1.541	3.826	1.122	0.740
40	1.862	3.849	1.181	0.717
45	2.183	3.843	1.242	0.723
50	2.506	3.872	1.299	0.695
55	2.828	3.889	1.358	0.678
60	3.144	3.825	1.422	0.741
65	3.467	3.882	1.479	0.684
70	3.791	3.835	1.537	0.732
75	4.118	3.889	1.590	0.677
80	4.445	3.873	1.644	0.694
85	4.791	4.070	1.678	0.496
90	5.128	4.114	1.722	0.452
95	5.462	4.023	1.769	0.543
100	5.803	4.041	1.809	0.525
105	6.149	4.064	1.842	0.502
110	6.495	4.124	1.878	0.442
115	6.821	3.992	1.932	0.575
120	7.164	4.002	1.969	0.564
125	7.525	4.228	1.989	0.338
130	7.858	4.132	2.037	0.435

OIL TYPE - ALAPAHA LOAMY SAND
DENTIFICATION CODE - 01012W
OVER - WEEDS-90, BARE-10
ATE OF RUN - 11 01 69
AINFALL INTENSITY - 2.884 INCHES/HOUR
NITIAL SOIL MOISTURE FOR THE 0 TO 12 INCH DEPTH - 3.48 INCHES
NITIAL SOIL MOISTURE FOR THE 12 TO 36 INCH DEPTH - 7.11 INCHES
INAL SOIL MOISTURE FOR THE 0 TO 12 INCH DEPTH - 3.53 INCHES
INAL SOIL MOISTURE FOR THE 12 TO 36 INCH DEPTH - 7.07 INCHES

TIME FROM TART OF RAIN (MINUTES)	ACCUMULATED RUNOFF (INCHES)	RUNOFF RATE (IN/HR)	ACCUMULATED INFILTRATION (INCHES)	INFILTRATION RATE (IN/HR)
4	0.000	0.000	0.192	2.884
5	0.012	0.961	0.228	1.923
10	0.168	2.183	0.312	0.700
15	0.364	2.349	0.356	0.534
20	0.556	2.368	0.405	0.515
25	0.763	2.511	0.438	0.372
30	0.971	2.523	0.471	0.361
35	1.180	2.546	0.502	0.338
40	1.388	2.527	0.534	0.356
45	1.602	2.547	0.560	0.336
50	1.816	2.572	0.587	0.312
55	2.026	2.543	0.617	0.340
60	2.240	2.557	0.643	0.327
65	2.445	2.498	0.679	0.385
70	2.661	2.515	0.703	0.369
75	2.874	2.493	0.731	0.391
80	3.086	2.494	0.759	0.390
85	3.299	2.510	0.787	0.373
90	3.509	2.483	0.817	0.401
95	3.721	2.487	0.845	0.397
100	3.939	2.525	0.867	0.358
105	4.149	2.534	0.897	0.349
110	4.357	2.480	0.931	0.404
115	4.571	2.516	0.956	0.367
120	4.788	2.578	0.980	0.305

CARNEGIE SANDY LOAM (02)

Location: 0.8 mi northwest of Engineering Building at Abraham Baldwin Agricultural College along field road; 100 ft west of road in pasture area; Tift County, Ga.

Land use or cover: Coastal bermudagrass.

Topography: Gently sloping — 6½%.

Great soil group: Fragic paleudults; fine-loamy, siliceous, thermic.

Parent material: Unconsolidated marine sediments of sandy clay loam.

Drainage: Well drained.

Horizon and Description

Apcn: 0 to 6 inches. Brown (10YR–4/3) sandy loam with some coarse sand grains; weak, fine granular structure; very friable; many small hard iron pebbles one-eighth to one-half inch in diameter; many fine roots; very strongly acid; abrupt smooth boundary.

B21tcn: 6 to 18 inches. Strong-brown (7.5YR–5/8) sandy clay loam; moderate, medium subangular blocky structure; friable, slightly sticky; iron pebbles common fine roots common; very strongly acid; gradual smooth boundary.

B22tp1: 18 to 34 inches. Yellowish-brown (10YR–5/(sandy clay) loam with common medium distinct mottle: of red (2.5YR–4/8), light gray (10YR–7/1), and yellowish red (5YR–4/8); moderate, medium subangula blocky structure; firm, slightly sticky; few roots; fev hard iron pebbles; soft plinthite; very strongly acid clear wavy boundary.

B23tp1: 34 to 60 inches. Reticulately mottled re((10YR–4/8), light gray (10YR–7/1), strong brow (7.5YR–5/8), and yellowish red (5YR–4/8) fine sand; clay loam, moderate, medium angular blocky structure firm, sticky, soft plinthite 15% to 30% by volume very strongly acid.

Remarks: Colors are given for moist soil. Reaction de termined by Soiltex.

WEIGHT PERCENT AND VOLUME PERCENT OF WATER RETAINED

DEPTH (inches)	TENSIONS (BARS)					BD G/CC	TP PCT	K
	.⊥	.3	.6	3.	15.			
0–6	8.38	7.52	7.46	7.31	6.34	1.61[1]	39.25	2.00–6.30
	13.49	12.11	12.01	11.77	10.21	1.63	38.49	
	FRAGMENT	7.22		SIEVED	5.05	ROCK PCT	21.30	
6–18	17.02	14.53	13.32	12.30	9.92	1.53[1]	42.26	0.63–2.00
	26.04	22.23	20.38	18.82	15.18	1.61	39.25	
	FRAGMENT	14.19		SIEVED	8.32	ROCK PCT	8.53	
18–34	18.71	18.18	15.43	11.62	6.00	1.64[1]	38.11	0.63–2.00
	30.68	29.82	25.31	19.06	9.84	1.67	36.98	
	FRAGMENT	16.78		SIEVED	7.66	ROCK PCT	7.58	
34+	17.91	17.03	13.04	9.37	4.85	1.63[1]	38.49	0.20–0.63
	29.19	27.76	21.26	15.27	7.91	1.71	35.47	
	FRAGMENT	16.35		SIEVED	5.59	ROCK PCT	13.34	

1=FIST
2=CORE
3=LOOSE

SOIL TYPE - CARNEGIE LOAMY SAND
IDENTIFICATION CODE - 02021D
COVER - GRASS-100
DATE OF RUN - 10 21 69
RAINFALL INTENSITY - 6.610 INCHES/HOUR
INITIAL SOIL MOISTURE FOR THE 0 TO 12 INCH DEPTH - 0.93 INCHES
INITIAL SOIL MOISTURE FOR THE 12 TO 36 INCH DEPTH - 5.43 INCHES
FINAL SOIL MOISTURE FOR THE 0 TO 12 INCH DEPTH - 3.33 INCHES
FINAL SOIL MOISTURE FOR THE 12 TO 36 INCH DEPTH - 6.78 INCHES

TIME FROM START OF RAIN (MINUTES)	ACCUMULATED RUNOFF (INCHES)	RUNOFF RATE (IN/HR)	ACCUMULATED INFILTRATION (INCHES)	INFILTRATION RATE (IN/HR)
7	0.000	0.000	0.771	6.610
10	0.076	3.485	0.901	3.124
15	0.541	3.576	1.111	3.034
20	0.842	3.528	1.361	3.081
25	1.131	3.457	1.622	3.152
30	1.415	3.372	1.890	3.237
35	1.692	3.315	2.163	3.294
40	1.964	3.265	2.442	3.345
45	2.236	3.220	2.721	3.389
50	2.494	3.026	3.014	3.584
55	2.735	2.860	3.323	3.750
60	2.968	2.759	3.642	3.851
65	3.181	2.474	3.979	4.136
70	3.394	2.445	4.318	4.165
75	3.596	2.456	4.667	4.153
80	3.798	2.458	5.015	4.151
85	3.997	2.367	5.367	4.242
90	4.197	2.372	5.717	4.238
95	4.408	2.394	6.058	4.215
100	4.610	2.451	6.407	4.159
105	4.835	2.734	6.732	3.876
110	5.065	2.911	7.054	3.699
115	5.320	3.098	7.349	3.512
120	5.577	3.180	7.643	3.429
125	5.872	3.427	7.899	3.183
130	6.153	3.383	8.169	3.227
135	6.437	3.416	8.436	3.193

SOIL TYPE - CARNEGIE LOAMY SAND
IDENTIFICATION CODE - 02021W
COVER - GRASS-100
DATE OF RUN - 10 21 69
RAINFALL INTENSITY - 5.528 INCHES/HOUR
INITIAL SOIL MOISTURE FOR THE 0 TO 12 INCH DEPTH - 2.67 INCHES
INITIAL SOIL MOISTURE FOR THE 12 TO 36 INCH DEPTH - 6.50 INCHES
FINAL SOIL MOISTURE FOR THE 0 TO 12 INCH DEPTH - 3.28 INCHES
FINAL SOIL MOISTURE FOR THE 12 TO 36 INCH DEPTH - 6.66 INCHES

TIME FROM START OF RAIN (MINUTES)	ACCUMULATED RUNOFF (INCHES)	RUNOFF RATE (IN/HR)	ACCUMULATED INFILTRATION (INCHES)	INFILTRATION RATE (IN/HR)
9	0.000	0.000	0.829	5.528
10	0.052	3.004	0.869	2.524
15	0.296	2.950	1.085	2.578
20	0.581	3.654	1.261	1.874
25	0.899	3.898	1.404	1.630
30	1.222	3.884	1.542	1.644
35	1.541	3.837	1.683	1.691
40	1.863	3.810	1.822	1.718
45	2.175	3.735	1.971	1.793
50	2.486	3.704	2.120	1.824
55	2.785	3.623	2.282	1.905
60	3.086	3.565	2.442	1.963
65	3.388	3.538	2.601	1.990
70	3.701	3.607	2.749	1.921
75	4.005	3.615	2.905	1.913
80	4.296	3.477	3.075	2.051
85	4.594	3.480	3.237	2.047
90	4.891	3.530	3.401	1.998
95	5.188	3.482	3.565	2.045
100	5.474	3.398	3.740	2.130
105	5.758	3.360	3.916	2.168
110	6.070	3.607	4.065	1.921
115	6.345	3.456	4.251	2.072
120	6.622	3.347	4.434	2.181
123	6.811	3.503	4.522	2.024

SOIL TYPE - CARNEGIE LOAMY SAND
IDENTIFICATION CODE - 02022D
COVER - GRASS-100
DATE OF RUN - 10 22 69
RAINFALL INTENSITY - 4.687 INCHES/HOUR
INITIAL SOIL MOISTURE FOR THE 0 TO 12 INCH DEPTH - 0.96 INCHES
INITIAL SOIL MOISTURE FOR THE 12 TO 36 INCH DEPTH - 5.93 INCHES
FINAL SOIL MOISTURE FOR THE 0 TO 12 INCH DEPTH - 3.31 INCHES
FINAL SOIL MOISTURE FOR THE 12 TO 36 INCH DEPTH - 7.11 INCHES

TIME FROM START OF RAIN (MINUTES)	ACCUMULATED RUNOFF (INCHES)	RUNOFF RATE (IN/HR)	ACCUMULATED INFILTRATION (INCHES)	INFILTRATION RATE (IN/HR)
4	0.000	0.000	0.312	4.687
5	0.028	1.682	0.362	3.004
10	0.160	1.669	0.620	3.017
15	0.276	1.394	0.895	3.292
20	0.395	1.428	1.167	3.259
25	0.516	1.435	1.437	3.252
30	0.639	1.470	1.704	3.216
35	0.758	1.459	1.976	3.228
40	0.880	1.431	2.244	3.256
45	0.991	1.359	2.524	3.328
50	1.102	1.344	2.804	3.343
55	1.214	1.373	3.082	3.313
60	1.330	1.380	3.357	3.306
65	1.445	1.503	3.632	3.184
70	1.583	1.656	3.885	3.030
75	1.722	1.664	4.137	3.022
80	1.855	1.773	4.394	2.914
85	2.015	2.063	4.624	2.623
90	2.191	2.131	4.839	2.556
95	2.366	2.129	5.054	2.558
100	2.548	2.182	5.263	2.504
105	2.726	2.133	5.476	2.553
110	2.907	2.126	5.685	2.560
115	3.091	2.179	5.892	2.507
120	3.265	2.097	6.108	2.589
125	3.447	2.093	6.318	2.594
130	3.631	2.157	6.524	2.530
135	3.805	2.108	6.741	2.578
140	3.988	2.126	6.949	2.560
145	4.168	2.120	7.158	2.567
150	4.345	2.090	7.373	2.597

SOIL TYPE — CARNEGIE LOAMY SAND
IDENTIFICATION CODE — 02022W
COVER — GRASS—100
DATE OF RUN — 10 22 69
RAINFALL INTENSITY — 3.004 INCHES/HOUR
INITIAL SOIL MOISTURE FOR THE 0 TO 12 INCH DEPTH — 2.76 INCHES
INITIAL SOIL MOISTURE FOR THE 12 TO 36 INCH DEPTH — 6.77 INCHES
FINAL SOIL MOISTURE FOR THE 0 TO 12 INCH DEPTH — 3.30 INCHES
FINAL SOIL MOISTURE FOR THE 12 TO 36 INCH DEPTH — 6.78 INCHES

TIME FROM START OF RAIN (MINUTES)	ACCUMULATED RUNOFF (INCHES)	RUNOFF RATE (IN/HR)	ACCUMULATED INFILTRATION (INCHES)	INFILTRATION RATE (IN/HR)
9	0.000	0.000	0.450	3.004
10	0.008	0.480	0.492	2.523
15	0.056	0.651	0.695	2.352
20	0.128	1.009	0.873	1.994
25	0.220	1.115	1.031	1.888
30	0.311	1.158	1.190	1.846
35	0.415	1.309	1.337	1.695
40	0.533	1.461	1.469	1.543
45	0.657	1.498	1.596	1.506
50	0.781	1.495	1.722	1.508
55	0.906	1.504	1.847	1.500
60	1.031	1.505	1.973	1.498
65	1.153	1.487	2.102	1.517
70	1.278	1.471	2.227	1.533
75	1.397	1.428	2.358	1.575
80	1.517	1.411	2.488	1.593
85	1.633	1.378	2.622	1.626
90	1.752	1.422	2.754	1.581
95	1.864	1.356	2.893	1.648
100	1.980	1.340	3.027	1.663
105	2.098	1.378	3.160	1.625
110	2.206	1.334	3.302	1.670

COWARTS LOAMY SAND (03, 16, and 17)

Location: 0.3 mi west of Animal Disease Laboratory along hard surface road to junction with U.S. 41; 270 yd northwest through pasture area to cultivated field; Tift County, Ga.

Land use or cover: Corn.

Topography: Very gently sloping — 3%.

Great soil group: Fragic paleudults; fine-loamy, siliceous, thermic.

Parent material: Unconsolidated marine sediments of sandy clay loam.

Drainage: Well drained.

Horizon and Description

Ap: 0 to 8 inches. Dark grayish-brown (10YR–4/2) loamy sand; weak fine granular structure; very friable, nonsticky; few small quartz gravel and common coarse sand grains; many fine roots; very strongly acid; abrupt smooth boundary.

A2: 8 to 12 inches. Pale brown (10YR–6/3) loamy sand; weak fine granular structure; very friable, nonsticky; common coarse sand grains; fine roots common; very strongly acid; clear smooth boundary.

B1t: to 15 inches. Light olive-brown (2.5YR–5/4) sandy loam; weak, medium granular structure; very friable, nonsticky; very strongly acid; clear wavy boundary.

B21t: 15 to 22 inches. Light olive-brown (2.5YR–5/6) sandy clay loam; moderate, medium subangular blocky structure; friable, slightly sticky; patchy clay films on peds; coarse sand grains coated and bridged with clay; very strongly acid; clear wavy boundary.

B22t: 22 to 39 inches. Yellowish-brown (10YR–5/6) fine sandy clay loam; common medium distinct and prominent mottles of red (2.5YR–4/8), light gray (10YR–7/1); brownish yellow (10YR–6/6) and yellowish red (5YR–5/8); moderate, medium subangular blocky structure; firm, slightly sticky; patchy clay films on ped faces; very strongly acid; gradual wavy boundary. This was horizon of least permeability.

B23tp1: 39 to 65 inches. Brownish-yellow (10YR–6/6) sandy clay loam with pockets of coarser and finer material; many coarse distinct and prominent mottles of light gray (10YR–7/1), red (2.5YR–4/8), and dusky red (7.5YR–3/4) moderate, medium subangular blocky and massive structure; firm, slightly sticky; soft plinthite 10% to 30% by volume; very strongly acid.

Remarks: Colors are given for moist soil. Reaction determined by Soiltex.

COWARTS LOAMY SAND (03 and 16 and 17)

WEIGHT PERCENT AND VOLUME PERCENT OF WATER RETAINED

DEPTH (inches)	TENSIONS (BARS)					BD G/CC	TP PCT	K
	.1	.3	.6	3.	15.			
0–5	10.24	5.96	3.67	3.62	2.33	1.67[1]	36.98	2.00–6.30
	17.10	9.95	6.13	6.05	3.89	1.66	37.36	
	FRAGMENT	5.69		SIEVED	2.52	ROCK PCT	6.39	
5–19	23.80	13.33	10.03	9.79	8.43	1.57[1]	40.75	0.63–2.00
	37.37	20.93	15.75	15.37	13.24	1.53	42.26	
	FRAGMENT	10.25		SIEVED	8.27	ROCK PCT	9.32	
19–32	23.47	18.62	13.61	12.34	11.37	1.55[1]	41.51	0.63–2.00
	36.38	28.86	21.10	19.13	17.62	1.56	41.13	
	FRAGMENT	17.37		SIEVED	11.25	ROCK PCT	10.07	
32+	25.96	15.13	13.67	12.18	11.56	1.53[1]	42.26	0.63–2.00
	39.72	23.15	20.92	18.64	17.69	1.55	41.51	
	FRAGMENT	14.06		SIEVED	10.86	ROCK PCT	5.21	

1=FIST
2=CORE
3=LOOSE

SOIL TYPE - COWARTS LOAMY SAND
IDENTIFICATION CODE - 03031D
COVER - WEEDS-80, BARE-20
DATE OF RUN - 10 17 69
RAINFALL INTENSITY - 5.168 INCHES/HOUR
INITIAL SOIL MOISTURE FOR THE 0 TO 12 INCH DEPTH - 1.93 INCHES
INITIAL SOIL MOISTURE FOR THE 12 TO 36 INCH DEPTH - 6.93 INCHES
FINAL SOIL MOISTURE FOR THE 0 TO 12 INCH DEPTH - 3.43 INCHES
FINAL SOIL MOISTURE FOR THE 12 TO 36 INCH DEPTH - 7.42 INCHES

TIME FROM START OF RAIN (MINUTES)	ACCUMULATED RUNOFF (INCHES)	RUNOFF RATE (IN/HR)	ACCUMULATED INFILTRATION (INCHES)	INFILTRATION RATE (IN/HR)
8	0.000	0.000	0.689	5.168
10	0.004	0.120	0.853	5.047
15	0.018	0.119	1.274	5.048
20	0.024	0.148	1.698	5.019
25	0.040	0.172	2.113	4.995
30	0.052	0.160	2.531	5.007
35	0.068	0.217	2.946	4.950
40	0.088	0.223	3.357	4.944
45	0.112	0.556	3.763	4.612
50	0.180	0.847	4.126	4.320
55	0.252	1.077	4.484	4.090
60	0.364	1.577	4.804	3.590
65	0.508	1.657	5.089	3.510
70	0.646	1.996	5.383	3.171
75	0.843	2.486	5.617	2.681
80	1.045	2.474	5.845	2.694
85	1.249	2.373	6.072	2.794
90	1.442	2.296	6.309	2.872
95	1.633	2.281	6.549	2.886
100	1.823	2.212	6.790	2.955
105	2.001	2.136	7.042	3.031
110	2.184	2.175	7.290	2.992
115	2.365	2.175	7.540	2.992
120	2.541	2.144	7.794	3.024
125	2.719	2.128	8.047	3.039
130	2.903	2.138	8.293	3.029

SOIL TYPE - COWARTS LOAMY SAND
IDENTIFICATION CODE - 03031W
COVER - WEEDS-80, BARE-20
DATE OF RUN - 10 17 69
RAINFALL INTENSITY - 6.009 INCHES/HOUR
INITIAL SOIL MOISTURE FOR THE 0 TO 12 INCH DEPTH - 3.06 INCHES
INITIAL SOIL MOISTURE FOR THE 12 TO 36 INCH DEPTH - 7.38 INCHES
FINAL SOIL MOISTURE FOR THE 0 TO 12 INCH DEPTH - 3.41 INCHES
FINAL SOIL MOISTURE FOR THE 12 TO 36 INCH DEPTH - 7.32 INCHES

TIME FROM START OF RAIN (MINUTES)	ACCUMULATED RUNOFF (INCHES)	RUNOFF RATE (IN/HR)	ACCUMULATED INFILTRATION (INCHES)	INFILTRATION RATE (IN/HR)
5	0.000	0.000	0.500	6.009
10	0.103	1.305	0.898	4.703
15	0.280	2.363	1.222	3.645
20	0.480	2.409	1.523	3.600
25	0.682	2.488	1.821	3.520
30	0.889	2.402	2.114	3.606
35	1.084	2.444	2.421	3.565
40	1.289	2.348	2.717	3.660
45	1.474	2.244	3.032	3.765
50	1.668	2.308	3.339	3.700
55	1.849	2.160	3.658	3.849
60	2.027	1.986	3.982	4.022
65	2.178	1.727	4.331	4.282
70	2.318	1.598	4.692	4.411
75	2.444	1.540	5.067	4.469
80	2.571	1.494	5.441	4.515
85	2.691	1.434	5.821	4.575
90	2.809	1.406	6.204	4.602
95	2.933	1.461	6.581	4.547
100	3.056	1.478	6.959	4.530
105	3.172	1.451	7.343	4.557
110	3.293	1.419	7.724	4.590
115	3.406	1.331	8.111	4.678
120	3.531	1.401	8.487	4.607

SOIL TYPE - COWARTS LOAMY SAND
IDENTIFICATION CODE - 03031WW
COVER - WEEDS-80, BARE-20
DATE OF RUN - 10 17 69
RAINFALL INTENSITY - 5.168 INCHES/HOUR
INITIAL SOIL MOISTURE FOR THE 0 TO 12 INCH DEPTH - 3.30 INCHES
INITIAL SOIL MOISTURE FOR THE 12 TO 36 INCH DEPTH - 7.11 INCHES
FINAL SOIL MOISTURE FOR THE 0 TO 12 INCH DEPTH - 3.17 INCHES
FINAL SOIL MOISTURE FOR THE 12 TO 36 INCH DEPTH - 7.02 INCHES

TIME FROM START OF RAIN (MINUTES)	ACCUMULATED RUNOFF (INCHES)	RUNOFF RATE (IN/HR)	ACCUMULATED INFILTRATION (INCHES)	INFILTRATION RATE (IN/HR)
2	0.000	0.000	0.172	5.168
5	0.008	0.600	0.390	4.567
10	0.100	0.901	0.761	4.267
15	0.180	1.080	1.111	4.087
20	0.280	1.283	1.442	3.884
25	0.399	1.785	1.753	3.382
30	0.576	2.196	2.007	2.971
35	0.757	2.172	2.257	2.995
40	0.938	2.181	2.506	2.987

SOIL TYPE - COWARTS LOAMY SAND
IDENTIFICATION CODE - 03032D
COVER - WEEDS-80, BARE-20
DATE OF RUN - 10 18 69
RAINFALL INTENSITY - 3.365 INCHES/HOUR
INITIAL SOIL MOISTURE FOR THE 0 TO 12 INCH DEPTH - 1.41 INCHES
INITIAL SOIL MOISTURE FOR THE 12 TO 36 INCH DEPTH - 6.17 INCHES
FINAL SOIL MOISTURE FOR THE 0 TO 12 INCH DEPTH - 3.29 INCHES
FINAL SOIL MOISTURE FOR THE 12 TO 36 INCH DEPTH - 6.78 INCHES

TIME FROM START OF RAIN (MINUTES)	ACCUMULATED RUNOFF (INCHES)	RUNOFF RATE (IN/HR)	ACCUMULATED INFILTRATION (INCHES)	INFILTRATION RATE (IN/HR)
30	0.000	0.000	1.682	3.365
35	0.004	0.048	1.959	3.317
40	0.008	0.048	2.235	3.317
45	0.012	0.048	2.512	3.317
50	0.016	0.047	2.788	3.317
55	0.020	0.047	3.064	3.317
60	0.024	0.047	3.341	3.318
65	0.028	0.044	3.617	3.321
70	0.032	0.072	3.894	3.293
75	0.040	0.101	4.166	3.263
80	0.048	0.097	4.438	3.267
85	0.056	0.097	4.711	3.267
90	0.064	0.096	4.983	3.268
95	0.072	0.095	5.256	3.269
100	0.080	0.095	5.528	3.269
105	0.088	0.094	5.801	3.271
110	0.096	0.094	6.073	3.270
115	0.104	0.109	6.345	3.255
120	0.114	0.117	6.616	3.247
125	0.124	0.125	6.887	3.239
130	0.136	0.171	7.155	3.193
135	0.152	0.200	7.419	3.164
140	0.168	0.193	7.684	3.171
145	0.184	0.191	7.948	3.173
150	0.200	0.188	8.213	3.176

SOIL TYPE - COWARTS LOAMY SAND
IDENTIFICATION CODE - 03032W
COVER - WEEDS-80, BARE-20
DATE OF RUN - 10 18 69
RAINFALL INTENSITY - 4.807 INCHES/HOUR
INITIAL SOIL MOISTURE FOR THE 0 TO 12 INCH DEPTH - 2.71 INCHES
INITIAL SOIL MOISTURE FOR THE 12 TO 36 INCH DEPTH - 6.60 INCHES
FINAL SOIL MOISTURE FOR THE 0 TO 12 INCH DEPTH - 3.40 INCHES
FINAL SOIL MOISTURE FOR THE 12 TO 36 INCH DEPTH - 6.67 INCHES

TIME FROM START OF RAIN (MINUTES)	ACCUMULATED RUNOFF (INCHES)	RUNOFF RATE (IN/HR)	ACCUMULATED INFILTRATION (INCHES)	INFILTRATIO RATE (IN/HR)
6	0.000	0.000	0.480	4.807
10	0.040	0.951	0.761	3.855
15	0.164	2.055	1.037	2.751
20	0.344	2.071	1.258	2.736
25	0.520	2.465	1.482	2.342
30	0.758	2.969	1.645	1.838
35	1.008	3.094	1.796	1.713
40	1.267	3.157	1.937	1.650
45	1.525	3.104	2.080	1.703
50	1.788	3.143	2.218	1.664
55	2.044	3.101	2.362	1.705
60	2.306	3.100	2.501	1.707
65	2.566	3.139	2.642	1.668
70	2.826	3.114	2.782	1.692
75	3.093	3.125	2.916	1.681
80	3.360	3.154	3.049	1.652
85	3.619	3.076	3.191	1.731
90	3.894	3.195	3.316	1.612

```
SOIL TYPE - COWARTS LOAMY SAND
IDENTIFICATION CODE - 03032WW
COVER - WEEDS-80, BARE-20
DATE OF RUN - 10 18 69
RAINFALL INTENSITY - 3.365 INCHES/HOUR
INITIAL SOIL MOISTURE FOR THE 0 TO 12 INCH DEPTH - 3.39 INCHES
INITIAL SOIL MOISTURE FOR THE 12 TO 36 INCH DEPTH - 6.71 INCHES
FINAL SOIL MOISTURE FOR THE 0 TO 12 INCH DEPTH -  3.50 INCHES
FINAL SOIL MOISTURE FOR THE 12 TO 36 INCH DEPTH -  6.81 INCHES
```

TIME FROM START OF RAIN (MINUTES)	ACCUMULATED RUNOFF (INCHES)	RUNOFF RATE (IN/HR)	ACCUMULATED INFILTRATION (INCHES)	INFILTRATION RATE (IN/HR)
8	0.000	0.000	0.112	3.365
10	0.008	0.721	0.540	2.644
15	0.100	1.210	0.741	2.155
20	0.200	1.244	0.921	2.121
25	0.304	1.248	1.098	2.116
30	0.407	1.230	1.275	2.134
35	0.512	1.253	1.450	2.111
40	0.619	1.279	1.624	2.085
45	0.722	1.261	1.801	2.104
50	0.826	1.261	1.978	2.104

SOIL TYPE - COWARTS LOAMY SAND
IDENTIFICATION CODE - 16031D
COVER - WEEDS-60, CORN-40
DATE OF RUN - 10 06 69
RAINFALL INTENSITY - 4.447 INCHES/HOUR
INITIAL SOIL MOISTURE FOR THE 0 TO 12 INCH DEPTH - 2.15 INCHES
INITIAL SOIL MOISTURE FOR THE 12 TO 36 INCH DEPTH - 6.47 INCHES
FINAL SOIL MOISTURE FOR THE 0 TO 12 INCH DEPTH - 3.22 INCHES
FINAL SOIL MOISTURE FOR THE 12 TO 36 INCH DEPTH - 7.09 INCHES

TIME FROM START OF RAIN (MINUTES)	ACCUMULATED RUNOFF (INCHES)	RUNOFF RATE (IN/HR)	ACCUMULATED INFILTRATION (INCHES)	INFILTRATION RATE (IN/HR)
4	0.000	0.000	0.296	4.447
5	0.020	1.081	0.350	3.365
10	0.096	1.052	0.645	3.394
15	0.176	0.863	0.935	3.583
20	0.240	0.772	1.242	3.674
25	0.308	0.819	1.544	3.627
30	0.376	0.796	1.846	3.650
35	0.440	0.743	2.153	3.703
40	0.501	0.719	2.463	3.728
45	0.561	0.716	2.774	3.730
50	0.621	0.768	3.084	3.679
55	0.690	0.884	3.385	3.562
60	0.765	0.952	3.681	3.494
65	0.845	0.968	3.971	3.478
70	0.926	0.974	4.261	3.472
75	1.003	0.943	4.555	3.503
80	1.084	0.928	4.844	3.518
85	1.162	0.894	5.137	3.553
90	1.233	0.832	5.437	3.614
95	1.301	0.776	5.739	3.670
100	1.366	0.794	6.044	3.652
105	1.435	0.837	6.346	3.609
110	1.502	0.812	6.650	3.634
115	1.569	0.841	6.954	3.605
120	1.647	0.959	7.246	3.487
125	1.727	0.973	7.537	3.473
130	1.806	0.957	7.829	3.489
135	1.888	0.987	8.117	3.459
140	1.966	0.970	8.409	3.476

SOIL TYPE - COWARTS LOAMY SAND
IDENTIFICATION CODE - 16031W
COVER - WEEDS-60, CORN-40
DATE OF RUN - 10 06 69
RAINFALL INTENSITY - 6.500 INCHES/HOUR
INITIAL SOIL MOISTURE FOR THE 0 TO 12 INCH DEPTH - 2.99 INCHES
INITIAL SOIL MOISTURE FOR THE 12 TO 36 INCH DEPTH - 6.80 INCHES
FINAL SOIL MOISTURE FOR THE 0 TO 12 INCH DEPTH - 3.09 INCHES
FINAL SOIL MOISTURE FOR THE 12 TO 36 INCH DEPTH - 6.81 INCHES

TIME FROM START OF RAIN (MINUTES)	ACCUMULATED RUNOFF (INCHES)	RUNOFF RATE (IN/HR)	ACCUMULATED INFILTRATION (INCHES)	INFILTRATION RATE (IN/HR)
3	0.000	0.000	0.325	6.500
5	0.048	3.605	0.441	2.895
10	0.430	3.952	0.653	2.548
15	0.749	4.040	0.875	2.460
20	1.140	4.183	1.026	2.316
25	1.485	4.129	1.223	2.371
30	1.828	4.102	1.422	2.398
35	2.168	4.065	1.623	2.435
40	2.497	3.959	1.836	2.541
45	2.830	3.966	2.044	2.534
50	3.161	3.896	2.255	2.604
55	3.489	3.894	2.469	2.606
60	3.818	3.903	2.682	2.597
65	4.147	3.888	2.895	2.611
70	4.472	3.911	3.112	2.589
75	4.801	3.906	3.324	2.594
80	5.129	3.919	3.538	2.581
85	5.457	3.868	3.751	2.631
90	5.785	3.915	3.966	2.584
95	6.093	3.708	4.199	2.792
100	6.404	3.615	4.430	2.885
105	6.722	3.657	4.654	2.843
110	7.023	3.581	4.894	2.918
115	7.340	3.762	5.119	2.737
120	7.642	3.617	5.359	2.882
125	7.943	3.656	5.599	2.844
130	8.246	3.680	5.838	2.820
135	8.541	3.633	6.084	2.867
140	8.841	3.603	6.326	2.897
145	9.145	3.663	6.564	2.836
147	9.267	3.692	6.658	2.808

SOIL TYPE - COWARTS LOAMY SAND
IDENTIFICATION CODE - 16032D
COVER - WEEDS-60, CORN-40
DATE OF RUN - 10 05 69
RAINFALL INTENSITY - 4.447 INCHES/HOUR
INITIAL SOIL MOISTURE FOR THE 0 TO 12 INCH DEPTH - 2.35 INCHES
INITIAL SOIL MOISTURE FOR THE 12 TO 36 INCH DEPTH - 7.30 INCHES
FINAL SOIL MOISTURE FOR THE 0 TO 12 INCH DEPTH - 3.33 INCHES
FINAL SOIL MOISTURE FOR THE 12 TO 36 INCH DEPTH - 7.84 INCHES

TIME FROM START OF RAIN (MINUTES)	ACCUMULATED RUNOFF (INCHES)	RUNOFF RATE (IN/HR)	ACCUMULATED INFILTRATION (INCHES)	INFILTRATION RATE (IN/HR)
15	0.000	0.000	1.111	4.447
20	0.004	0.048	1.478	4.398
25	0.008	0.048	1.844	4.398
30	0.012	0.048	2.211	4.399
35	0.016	0.047	2.578	4.399
40	0.020	0.047	2.944	4.399
45	0.024	0.045	3.311	4.401
50	0.028	0.058	3.677	4.388
55	0.034	0.085	4.042	4.361
60	0.042	0.098	4.404	4.348
65	0.050	0.096	4.767	4.351
70	0.058	0.108	5.130	4.338
75	0.068	0.122	5.490	4.324
80	0.078	0.121	5.851	4.325
85	0.088	0.103	6.211	4.343
90	0.096	0.119	6.574	4.327
95	0.108	0.150	6.932	4.297
100	0.120	0.144	7.291	4.302
105	0.132	0.137	7.650	4.310
110	0.144	0.168	8.008	4.278
115	0.160	0.196	8.363	4.250
120	0.176	0.192	8.717	4.254
125	0.191	0.187	9.072	4.259
130	0.207	0.183	9.427	4.263
135	0.224	0.186	9.781	4.260
140	0.240	0.214	10.135	4.232
145	0.260	0.239	10.486	4.207
150	0.279	0.230	10.837	4.216

SOIL TYPE — COWARTS LOAMY SAND
IDENTIFICATION CODE — 16032W
COVER — WEEDS-60, CORN-40
DATE OF RUN — 10 05 69
RAINFALL INTENSITY — 6.500 INCHES/HOUR
INITIAL SOIL MOISTURE FOR THE 0 TO 12 INCH DEPTH — 3.35 INCHES
INITIAL SOIL MOISTURE FOR THE 12 TO 36 INCH DEPTH — 9.45 INCHES
FINAL SOIL MOISTURE FOR THE 0 TO 12 INCH DEPTH — 4.07 INCHES
FINAL SOIL MOISTURE FOR THE 12 TO 36 INCH DEPTH — 9.98 INCHES

TIME FROM START OF RAIN (MINUTES)	ACCUMULATED RUNOFF (INCHES)	RUNOFF RATE (IN/HR)	ACCUMULATED INFILTRATION (INCHES)	INFILTRATION RATE (IN/HR)
2	0.000	0.000	0.216	6.500
5	0.012	1.562	0.469	4.938
10	0.240	2.508	0.843	3.992
15	0.480	2.720	1.144	3.780
20	0.681	2.378	1.485	4.121
25	0.882	2.399	1.825	4.101
30	1.079	2.370	2.170	4.130
35	1.273	2.369	2.519	4.130
40	1.473	2.393	2.860	4.106
45	1.677	2.463	3.198	4.037
50	1.881	2.536	3.535	3.964
55	2.088	2.600	3.870	3.900
60	2.316	2.720	4.184	3.780
65	2.542	2.719	4.500	3.781
70	2.767	2.704	4.816	3.796
75	2.999	2.737	5.126	3.763
80	3.219	2.610	5.447	3.890
85	3.454	2.675	5.755	3.824
90	3.682	2.688	6.068	3.812
95	3.923	2.827	6.369	3.673
100	4.149	2.695	6.684	3.805
105	4.396	2.897	6.979	3.603
110	4.640	2.962	7.277	3.537
115	4.887	2.986	7.572	3.513
120	5.134	3.006	7.867	3.494
125	5.355	2.746	8.187	3.754
130	5.619	2.970	8.465	3.529
135	5.877	3.142	8.748	3.358

SOIL TYPE - COWARTS LOAMY SAND
IDENTIFICATION CODE - 16033D
COVER - WEEDS-60, CORN-40
DATE OF RUN - 10 04 69
RAINFALL INTENSITY - 4.326 INCHES/HOUR
INITIAL SOIL MOISTURE FOR THE O TO 12 INCH DEPTH - 2.28 INCHES
INITIAL SOIL MOISTURE FOR THE 12 TO 36 INCH DEPTH - 7.75 INCHES
FINAL SOIL MOISTURE FOR THE O TO 12 INCH DEPTH - 3.96 INCHES
FINAL SOIL MOISTURE FOR THE 12 TO 36 INCH DEPTH - 8.28 INCHES

TIME FROM START OF RAIN (MINUTES)	ACCUMULATED RUNOFF (INCHES)	RUNOFF RATE (IN/HR)	ACCUMULATED INFILTRATION (INCHES)	INFILTRATIC RATE (IN/HR)
7	0.000	0.000	0.504	4.326
10	0.002	0.120	0.715	4.206
15	0.016	0.120	1.065	4.206
20	0.030	0.124	1.412	4.202
25	0.039	0.094	1.763	4.232
30	0.047	0.097	2.116	4.229
35	0.055	0.096	2.468	4.230
40	0.063	0.096	2.821	4.230
45	0.071	0.096	3.173	4.230
50	0.079	0.096	3.526	4.230
55	0.087	0.096	3.878	4.230
60	0.095	0.095	4.231	4.231
65	0.103	0.096	4.584	4.230
70	0.111	0.096	4.936	4.230
75	0.119	0.095	5.289	4.231
80	0.127	0.091	5.641	4.235
85	0.136	0.122	5.993	4.204
90	0.146	0.115	6.344	4.211
95	0.156	0.146	6.693	4.180
100	0.169	0.144	7.042	4.182
105	0.180	0.136	7.391	4.190
110	0.193	0.169	7.738	4.157
115	0.207	0.165	8.085	4.161
120	0.222	0.182	8.431	4.144
125	0.240	0.246	8.773	4.080
130	0.260	0.228	9.114	4.098
135	0.281	0.290	9.453	4.036
140	0.306	0.289	9.789	4.037
145	0.329	0.283	10.126	4.043
150	0.354	0.287	10.463	4.039
155	0.378	0.289	10.799	4.037
160	0.402	0.289	11.135	4.037
163	0.416	0.289	11.337	4.036

SOIL TYPE - COWARTS LOAMY SAND
IDENTIFICATION CODE - 16033W
COVER - WEEDS-60, CORN-40
DATE OF RUN - 10 04 69
RAINFALL INTENSITY - 2.884 INCHES/HOUR
INITIAL SOIL MOISTURE FOR THE 0 TO 12 INCH DEPTH - 3.58 INCHES
INITIAL SOIL MOISTURE FOR THE 12 TO 36 INCH DEPTH - 8.41 INCHES
FINAL SOIL MOISTURE FOR THE 0 TO 12 INCH DEPTH - 3.69 INCHES
FINAL SOIL MOISTURE FOR THE 12 TO 36 INCH DEPTH - 7.87 INCHES

TIME FROM START OF RAIN (MINUTES)	ACCUMULATED RUNOFF (INCHES)	RUNOFF RATE (IN/HR)	ACCUMULATED INFILTRATION (INCHES)	INFILTRATION RATE (IN/HR)
4	0.000	0.000	0.192	2.884
5	0.008	0.600	0.232	2.283
10	0.072	1.144	0.408	1.739
15	0.160	1.246	0.560	1.638
20	0.276	1.430	0.684	1.454
25	0.394	1.516	0.807	1.367
30	0.532	1.674	0.909	1.210
35	0.675	1.707	1.007	1.177
40	0.815	1.724	1.107	1.159
45	0.958	1.765	1.205	1.119
50	1.105	1.783	1.298	1.100
55	1.254	1.774	1.390	1.110
60	1.402	1.792	1.482	1.092
65	1.554	1.879	1.570	1.005
70	1.718	1.978	1.647	0.906
75	1.874	1.919	1.731	0.964
80	2.032	1.881	1.813	1.003
85	2.195	1.900	1.891	0.983
90	2.359	1.946	1.967	0.938
95	2.513	1.912	2.053	0.971
100	2.678	1.959	2.128	0.925
105	2.836	1.954	2.211	0.929
110	3.001	1.993	2.286	0.891
115	3.156	1.874	2.372	1.010
120	3.328	1.939	2.440	0.945

SOIL TYPE - COWARTS LOAMY SAND
IDENTIFICATION CODE - 17031D
COVER - GRASS-100
DATE OF RUN - 10 07 69
RAINFALL INTENSITY - 4.376 INCHES/HOUR
INITIAL SOIL MOISTURE FOR THE 0 TO 12 INCH DEPTH - 1.11 INCHES
INITIAL SOIL MOISTURE FOR THE 12 TO 36 INCH DEPTH - 5.85 INCHES
FINAL SOIL MOISTURE FOR THE 0 TO 12 INCH DEPTH - 3.06 INCHES
FINAL SOIL MOISTURE FOR THE 12 TO 36 INCH DEPTH - 6.67 INCHES

TIME FROM START OF RAIN (MINUTES)	ACCUMULATED RUNOFF (INCHES)	RUNOFF RATE (IN/HR)	ACCUMULATED INFILTRATION (INCHES)	INFILTRATION RATE (IN/HR)
4	0.000	0.000	0.282	4.326
5	0.004	0.480	0.356	3.846
10	0.052	0.471	0.669	3.855
15	0.092	0.453	0.989	3.873
20	0.128	0.410	1.314	3.916
25	0.160	0.362	1.642	3.964
30	0.187	0.308	1.975	4.018
35	0.212	0.262	2.311	4.064
40	0.232	0.215	2.652	4.111
45	0.248	0.166	2.996	4.160
50	0.260	0.135	3.345	4.191
55	0.272	0.166	3.693	4.160
60	0.288	0.196	4.038	4.130
65	0.304	0.194	4.382	4.132
70	0.320	0.187	4.727	4.139
75	0.336	0.190	5.072	4.136
80	0.352	0.195	5.416	4.131
85	0.369	0.197	5.760	4.129
90	0.384	0.183	6.105	4.143
95	0.401	0.222	6.449	4.104
100	0.421	0.248	6.790	4.078
105	0.441	0.243	7.130	4.083
110	0.460	0.234	7.472	4.092
115	0.481	0.247	7.811	4.079
120	0.502	0.256	8.151	4.069
125	0.520	0.234	8.493	4.092
130	0.540	0.236	8.834	4.090
135	0.561	0.243	9.174	4.083
140	0.581	0.249	9.514	4.077
145	0.600	0.239	9.855	4.087
150	0.622	0.257	10.194	4.069

SOIL TYPE - COWARTS LOAMY SAND
IDENTIFICATION CODE - 17031W
COVER - GRASS-100
DATE OF RUN - 10 07 69
RAINFALL INTENSITY - 6.249 INCHES/HOUR
INITIAL SOIL MOISTURE FOR THE 0 TO 12 INCH DEPTH - 2.44 INCHES
INITIAL SOIL MOISTURE FOR THE 12 TO 36 INCH DEPTH - 6.47 INCHES
FINAL SOIL MOISTURE FOR THE 0 TO 12 INCH DEPTH - 3.22 INCHES
FINAL SOIL MOISTURE FOR THE 12 TO 36 INCH DEPTH - 6.53 INCHES

TIME FROM START OF RAIN (MINUTES)	ACCUMULATED RUNOFF (INCHES)	RUNOFF RATE (IN/HR)	ACCUMULATED INFILTRATION (INCHES)	INFILTRATION RATE (IN/HR)
8	0.000	0.000	0.833	6.249
10	0.028	1.562	0.985	4.687
15	0.200	2.601	1.361	3.648
20	0.432	2.880	1.650	3.369
25	0.681	3.187	1.922	3.062
30	0.962	3.509	2.162	2.740
35	1.262	3.747	2.383	2.502
40	1.581	3.840	2.584	2.409
45	1.904	3.933	2.782	2.316
50	2.238	4.072	2.970	2.177
55	2.577	4.109	3.151	2.140
60	2.913	4.037	3.336	2.212
65	3.255	4.038	3.515	2.211
70	3.590	4.004	3.700	2.245
75	3.940	4.087	3.871	2.161
80	4.274	4.010	4.058	2.239
85	4.612	4.022	4.241	2.227
90	4.958	4.097	4.416	2.152
95	5.312	4.201	4.583	2.048
100	5.616	3.880	4.800	2.369
105	5.990	4.222	4.946	2.027
110	6.342	4.230	5.115	2.019
115	6.692	4.254	5.286	1.995
120	7.032	4.124	5.467	2.125

SOIL TYPE — COWARTS LOAMY SAND
IDENTIFICATION CODE — 17032D
COVER — GRASS-100
DATE OF RUN — 10 08 69
RAINFALL INTENSITY — 4.567 INCHES/HOUR
INITIAL SOIL MOISTURE FOR THE 0 TO 12 INCH DEPTH — 0.97 INCHES
INITIAL SOIL MOISTURE FOR THE 12 TO 36 INCH DEPTH — 5.53 INCHES
FINAL SOIL MOISTURE FOR THE 0 TO 12 INCH DEPTH — 3.64 INCHES
FINAL SOIL MOISTURE FOR THE 12 TO 36 INCH DEPTH — 6.77 INCHES

TIME FROM START OF RAIN (MINUTES)	ACCUMULATED RUNOFF (INCHES)	RUNOFF RATE (IN/HR)	ACCUMULATED INFILTRATION (INCHES)	INFILTRATION RATE (IN/HR)
7	0.000	0.000	0.532	4.567
10	0.008	0.961	0.721	3.605
15	0.120	1.031	1.021	3.535
20	0.212	1.024	1.310	3.542
25	0.288	0.873	1.614	3.693
30	0.359	0.826	1.924	3.740
35	0.428	0.790	2.235	3.777
40	0.493	0.750	2.551	3.817
45	0.552	0.699	2.872	3.868
50	0.609	0.630	3.196	3.937
55	0.657	0.557	3.529	4.010
60	0.702	0.511	3.864	4.056
65	0.742	0.515	4.205	4.051
70	0.786	0.526	4.542	4.040
75	0.829	0.624	4.879	3.942
80	0.891	0.779	5.198	3.788
85	0.958	0.980	5.511	3.586
90	1.048	1.112	5.802	3.454
95	1.141	1.127	6.090	3.439
100	1.242	1.319	6.369	3.247
105	1.361	1.517	6.630	3.049
110	1.494	1.644	6.878	2.922
115	1.633	1.720	7.119	2.847
120	1.781	1.813	7.352	2.754
125	1.936	1.913	7.578	2.653
130	2.103	1.974	7.792	2.593
135	2.273	2.028	8.002	2.538
140	2.430	1.959	8.226	2.607
145	2.597	1.996	8.440	2.570
150	2.763	2.009	8.655	2.558
155	2.920	1.933	8.877	2.634

```
SOIL TYPE - COWARTS LOAMY SAND
IDENTIFICATION CODE - 17032W
COVER - GRASS-100
DATE OF RUN - 10 08 69
RAINFALL INTENSITY - 6.249 INCHES/HOUR
INITIAL SOIL MOISTURE FOR THE 0 TO 12 INCH DEPTH - 2.70 INCHES
INITIAL SOIL MOISTURE FOR THE 12 TO 36 INCH DEPTH - 6.61 INCHES
FINAL SOIL MOISTURE FOR THE 0 TO 12 INCH DEPTH -  3.59 INCHES
FINAL SOIL MOISTURE FOR THE 12 TO 36 INCH DEPTH -  6.49 INCHES
```

TIME FROM START OF RAIN (MINUTES)	ACCUMULATED RUNOFF (INCHES)	RUNOFF RATE (IN/HR)	ACCUMULATED INFILTRATION (INCHES)	INFILTRATION RATE (IN/HR)
6	0.000	0.000	0.625	6.249
10	0.040	1.481	1.001	4.767
15	0.291	2.931	1.271	3.318
20	0.641	4.678	1.441	1.571
25	1.043	4.868	1.560	1.381
30	1.443	4.796	1.681	1.453
35	1.845	4.920	1.800	1.329
40	2.260	5.029	1.905	1.220
45	2.664	4.924	2.022	1.325
50	3.082	4.942	2.126	1.307
55	3.491	4.870	2.237	1.379
60	3.900	4.822	2.349	1.427
65	4.304	4.761	2.465	1.488
70	4.700	4.685	2.591	1.563
75	5.105	4.791	2.706	1.458
80	5.498	4.707	2.834	1.542
85	5.901	4.750	2.952	1.498
90	6.312	4.822	3.062	1.427
95	6.705	4.783	3.190	1.466
100	7.126	4.999	3.289	1.250
105	7.530	5.011	3.406	1.238
110	7.956	5.119	3.501	1.130
115	8.366	4.978	3.612	1.271
120	8.804	5.146	3.695	1.102
125	9.221	5.122	3.799	1.127
130	9.660	5.357	3.880	0.892
135	10.070	5.203	3.991	1.046
137	10.232	5.203	4.037	1.046

SOIL TYPE - COWARTS LOAMY SAND
IDENTIFICATION CODE - 17033D
COVER - GRASS-100
DATE OF RUN - 10 09 69
RAINFALL INTENSITY - 2.764 INCHES/HOUR
INITIAL SOIL MOISTURE FOR THE 0 TO 12 INCH DEPTH - 1.02 INCHES
INITIAL SOIL MOISTURE FOR THE 12 TO 36 INCH DEPTH - 5.42 INCHES
FINAL SOIL MOISTURE FOR THE 0 TO 12 INCH DEPTH - 2.61 INCHES
FINAL SOIL MOISTURE FOR THE 12 TO 36 INCH DEPTH - 6.53 INCHES

TIME FROM START OF RAIN (MINUTES)	ACCUMULATED RUNOFF (INCHES)	RUNOFF RATE (IN/HR)	ACCUMULATED INFILTRATION (INCHES)	INFILTRATION RATE (IN/HR)
4	0.000	0.000	0.184	2.764
5	0.004	0.600	0.226	2.163
10	0.080	1.011	0.380	1.753
15	0.196	1.283	0.494	1.481
20	0.287	1.113	0.633	1.651
25	0.383	1.052	0.768	1.711
30	0.463	0.900	0.918	1.863
35	0.537	0.808	1.075	1.955
40	0.598	0.614	1.244	2.149
45	0.642	0.632	1.430	2.132
50	0.699	0.563	1.604	2.201
55	0.730	0.412	1.803	2.351
60	0.766	0.397	1.998	2.366
65	0.792	0.317	2.202	2.446
70	0.821	0.320	2.403	2.443
75	0.846	0.297	2.609	2.467
80	0.869	0.264	2.816	2.500
85	0.889	0.236	3.027	2.527
90	0.908	0.204	3.237	2.559
95	0.924	0.180	3.452	2.583
100	0.940	0.186	3.666	2.578
105	0.958	0.174	3.879	2.590
110	0.969	0.144	4.098	2.619
115	0.981	0.142	4.316	2.622
120	0.993	0.148	4.534	2.616
125	1.006	0.153	4.752	2.611
130	1.017	0.139	4.972	2.625

SOIL TYPE — COWARTS LOAMY SAND
IDENTIFICATION CODE — 17033W
COVER — GRASS—100
DATE OF RUN — 10 09 69
RAINFALL INTENSITY — 5.288 INCHES/HOUR
INITIAL SOIL MOISTURE FOR THE 0 TO 12 INCH DEPTH — 2.24 INCHES
INITIAL SOIL MOISTURE FOR THE 12 TO 36 INCH DEPTH — 6.41 INCHES
FINAL SOIL MOISTURE FOR THE 0 TO 12 INCH DEPTH — 2.84 INCHES
FINAL SOIL MOISTURE FOR THE 12 TO 36 INCH DEPTH — 6.63 INCHES

TIME FROM START OF RAIN (MINUTES)	ACCUMULATED RUNOFF (INCHES)	RUNOFF RATE (IN/HR)	ACCUMULATED INFILTRATION (INCHES)	INFILTRATION RATE (IN/HR)
6	0.000	0.000	0.528	5.288
10	0.140	2.929	0.741	2.359
15	0.378	2.743	0.943	2.545
20	0.597	2.658	1.165	2.629
25	0.819	2.671	1.384	2.617
30	1.039	2.639	1.604	2.648
35	1.254	2.593	1.830	2.694
40	1.469	2.587	2.055	2.701
45	1.688	2.624	2.278	2.663
50	1.902	2.543	2.504	2.744
55	2.109	2.464	2.738	2.824
60	2.324	2.534	2.964	2.753
65	2.526	2.482	3.202	2.805
70	2.734	2.506	3.435	2.781
75	2.945	2.437	3.664	2.851
80	3.143	2.368	3.907	2.920
85	3.349	2.429	4.142	2.858
90	3.541	2.353	4.391	2.935
95	3.738	2.296	4.634	2.991
100	3.943	2.346	4.870	2.942

SOIL TYPE — COWARTS LOAMY SAND
IDENTIFICATION CODE — 17033WW
COVER — GRASS—100
DATE OF RUN — 10 09 69
RAINFALL INTENSITY — 2.764 INCHES/HOUR
INITIAL SOIL MOISTURE FOR THE 0 TO 12 INCH DEPTH — 2.76 INCHES
INITIAL SOIL MOISTURE FOR THE 12 TO 36 INCH DEPTH — 6.52 INCHES
FINAL SOIL MOISTURE FOR THE 0 TO 12 INCH DEPTH — 2.38 INCHES
FINAL SOIL MOISTURE FOR THE 12 TO 36 INCH DEPTH — 6.42 INCHES

TIME FROM START OF RAIN (MINUTES)	ACCUMULATED RUNOFF (INCHES)	RUNOFF RATE (IN/HR)	ACCUMULATED INFILTRATION (INCHES)	INFILTRATION RATE (IN/HR)
8	0.000	0.000	0.368	2.764
10	0.002	0.132	0.456	2.632
15	0.015	0.161	0.675	2.603
20	0.028	0.192	0.893	2.572
25	0.048	0.248	1.103	2.516
30	0.068	0.240	1.314	2.523
35	0.088	0.239	1.524	2.525
40	0.108	0.242	1.734	2.522
45	0.128	0.241	1.945	2.522
50	0.148	0.241	2.155	2.522
55	0.168	0.239	2.365	2.524
60	0.188	0.239	2.576	2.524
65	0.207	0.234	2.786	2.529
70	0.228	0.239	2.996	2.524

DOTHAN LOAMY SAND (04)

Location: 1.5 mi north of Coastal Plain Experiment Station dairy barn along field road; west along field road for 0.6 mi; 15 ft south of road; Tift County, Ga.

Land use or cover: Corn.

Topography: Very gently sloping — 3%.

Great soil group: Plinthic paleudults; fine-loamy, siliceous, thermic.

Parent material: Unconsolidated marine sediments of sandy clay loam.

Drainage: Well drained.

Horizon and Description

Ap: 0 to 9 inches. Grayish-brown (10YR-5/2) loamy sand; weak, fine granular structure; very friable, nonsticky; few small hard iron pebbles; many fine roots; very strongly acid; abrupt smooth boundary.

B1t: 9 to 16 inches. Yellowish-brown (10YR-5/6) sandy loam and light sandy clay loam; weak, medium subangular blocky structure; friable; few small hard iron pebbles; fine roots common; very strongly acid; clear smooth boundary.

B21t: 16 to 36 inches. Brownish-yellow (10YR-6/6) sandy clay loam; moderate, medium subangular blocky structure; friable, slightly sticky; few roots; few small hard iron pebbles; very strongly acid; gradual wavy boundary.

B22t: 36 to 52 inches. Brownish-yellow (10YR-6/6) sandy loam with common medium distinct mottles of yellowish red (5YR-5/6), yellowish brown (10YR-5/6), and light gray (10YR-7/2); moderate, medium subangular blocky structure; friable, slightly sticky; few small hard iron pebbles; very strongly acid; gradual wavy boundary.

B23tp1: 52 to 65 inches. Light-yellowish-brown (2.5YR-6/4) sandy clay loam with many coarse distinct and prominent mottles of red (2.5YR-4/8), light gray (10YR-7/1) and yellowish brown (10YR-5/8); moderate, medium subangular blocky structure; firm, slightly sticky; soft plinthite 10% to 30% by volume; very strongly acid.

Remarks: Colors are given for moist soil. Reaction determined by Soiltex.

DOTHAN LOAMY SAND (04)

WEIGHT PERCENT AND VOLUME PERCENT OF WATER RETAINED

DEPTH (inches)	TENSIONS (BARS)					BD G/CC	TP PCT	K
	.1	.3	.6	3.	15.			
0-9	11.17	6.71	5.97	4.18	1.38	1.57[1]	40.75	2.00-6.30
	17.54	10.53	9.37	6.56	2.17	1.59	40.00	
	FRAGMENT	4.93		SIEVED	1.60	ROCK PCT	3.83	
9-16	12.55	9.77	8.77	7.14	6.39	1.58[1]	40.38	2.00-6.30
	19.83	15.44	13.86	11.28	10.10	1.66	37.36	
	FRAGMENT	10.43		SIEVED	6.63	ROCK PCT	6.56	
16-36	20.51	16.84	15.45	12.85	12.68	1.55[1]	41.51	0.63-2.00
	31.79	26.10	23.95	19.92	19.65	1.59	40.00	
	FRAGMENT	15.33		SIEVED	12.41	ROCK PCT	14.81	
36+	13.57	12.38	11.25	10.10	8.55	1.55[1]	41.51	0.63-2.00
	21.03	19.19	17.44	15.65	13.25	1.59	40.00	
	FRAGMENT	10.84		SIEVED	8.30	ROCK PCT	1.48	
52	12.63	11.29	10.40	9.15	3.33	1.57[1]	40.75	0.63-2.00
	19.83	17.73	16.33	14.37	5.23	1.66	37.36	
	FRAGMENT	9.81		SIEVED	2.10	ROCK PCT	1.32	

1=FIST
2=CORE
3=LOOSE

SOIL TYPE — DOTHAN LOAMY SAND
IDENTIFICATION CODE — 04041D
COVER — BARE-80, WEEDS-20
DATE OF RUN — 11 04 69
RAINFALL INTENSITY — 4.687 INCHES/HOUR
INITIAL SOIL MOISTURE FOR THE 0 TO 12 INCH DEPTH — 0.79 INCHES
INITIAL SOIL MOISTURE FOR THE 12 TO 36 INCH DEPTH — 4.04 INCHES
FINAL SOIL MOISTURE FOR THE 0 TO 12 INCH DEPTH — 3.57 INCHES
FINAL SOIL MOISTURE FOR THE 12 TO 36 INCH DEPTH — 6.97 INCHES

TIME FROM START OF RAIN (MINUTES)	ACCUMULATED RUNOFF (INCHES)	RUNOFF RATE (IN/HR)	ACCUMULATED INFILTRATION (INCHES)	INFILTRATION RATE (IN/HR)
10	0.000	0.000	0.781	4.687
15	0.006	0.078	1.165	4.608
20	0.016	0.120	1.546	4.567
25	0.024	0.120	1.929	4.567
30	0.036	0.148	2.307	4.538
35	0.048	0.136	2.686	4.550
40	0.060	0.188	3.064	4.498
45	0.080	0.260	3.435	4.426
50	0.104	0.362	3.801	4.324
55	0.140	0.458	4.156	4.228
60	0.180	0.534	4.507	4.152
65	0.228	0.558	4.849	4.129
70	0.272	0.537	5.195	4.150
75	0.319	0.643	5.539	4.044
80	0.380	0.725	5.869	3.961
85	0.440	0.741	6.199	3.946
90	0.504	0.798	6.526	3.888
95	0.572	0.816	6.848	3.871
100	0.641	0.856	7.170	3.830
105	0.713	0.837	7.489	3.849
110	0.782	0.889	7.811	3.797
115	0.861	1.022	8.122	3.665
120	0.950	1.098	8.424	3.589
125	1.040	1.068	8.724	3.618
130	1.130	1.087	9.025	3.600
135	1.221	1.079	9.324	3.607
140	1.309	1.067	9.627	3.619
145	1.401	1.098	9.926	3.589
150	1.494	1.083	10.224	3.603
155	1.582	1.027	10.526	3.659
160	1.669	1.033	10.830	3.654
165	1.743	0.857	11.146	3.829
170	1.817	0.837	11.463	3.849
175	1.890	0.857	11.781	3.829
180	1.965	0.896	12.096	3.790

SOIL TYPE - DOTHAN LOAMY SAND
IDENTIFICATION CODE - 04041W
COVER - BARE-80, WEEDS-20
DATE OF RUN - 11 04 69
RAINFALL INTENSITY - 3.365 INCHES/HOUR
INITIAL SOIL MOISTURE FOR THE 0 TO 12 INCH DEPTH - 2.87 INCHES
INITIAL SOIL MOISTURE FOR THE 12 TO 36 INCH DEPTH - 6.43 INCHES
FINAL SOIL MOISTURE FOR THE 0 TO 12 INCH DEPTH - 3.09 INCHES
FINAL SOIL MOISTURE FOR THE 12 TO 36 INCH DEPTH - 6.81 INCHES

TIME FROM START OF RAIN (MINUTES)	ACCUMULATED RUNOFF (INCHES)	RUNOFF RATE (IN/HR)	ACCUMULATED INFILTRATION (INCHES)	INFILTRATION RATE (IN/HR)
6	0.000	0.000	0.336	3.365
10	0.008	0.118	0.552	3.247
15	0.019	0.154	0.822	3.210
20	0.032	0.169	1.089	3.196
25	0.048	0.190	1.354	3.175
30	0.064	0.233	1.618	3.132
35	0.088	0.335	1.875	3.030
40	0.120	0.438	2.123	2.926
45	0.160	0.511	2.363	2.853
50	0.204	0.531	2.600	2.834
55	0.248	0.519	2.836	2.845
60	0.292	0.550	3.072	2.815
65	0.340	0.580	3.305	2.785
70	0.388	0.571	3.537	2.793
75	0.436	0.599	3.770	2.765
80	0.490	0.647	3.996	2.717
85	0.540	0.622	4.226	2.743
90	0.593	0.632	4.454	2.732
95	0.646	0.643	4.681	2.721
100	0.699	0.650	4.909	2.715
105	0.748	0.622	5.140	2.742
110	0.802	0.639	5.367	2.725
115	0.853	0.624	5.596	2.741
120	0.907	0.651	5.822	2.713

SOIL TYPE - DOTHAN LOAMY SAND
IDENTIFICATION CODE - 04042D
COVER - BARE-80, WEEDS-20
DATE OF RUN - 11 05 69
RAINFALL INTENSITY - 4.807 INCHES/HOUR
INITIAL SOIL MOISTURE FOR THE 0 TO 12 INCH DEPTH - 1.50 INCHES
INITIAL SOIL MOISTURE FOR THE 12 TO 36 INCH DEPTH - 5.35 INCHES
FINAL SOIL MOISTURE FOR THE 0 TO 12 INCH DEPTH - 3.01 INCHES
FINAL SOIL MOISTURE FOR THE 12 TO 36 INCH DEPTH - 7.01 INCHES

TIME FROM START OF RAIN (MINUTES)	ACCUMULATED RUNOFF (INCHES)	RUNOFF RATE (IN/HR)	ACCUMULATED INFILTRATION (INCHES)	INFILTRATION RATE (IN/HR)
5	0.000	0.000	0.400	4.807
10	0.020	0.238	0.781	4.568
15	0.040	0.231	1.161	4.575
20	0.060	0.240	1.542	4.566
25	0.080	0.241	1.922	4.566
30	0.100	0.239	2.303	4.568
35	0.120	0.239	2.684	4.568
40	0.140	0.241	3.064	4.565
45	0.164	0.382	3.441	4.425
50	0.207	0.701	3.798	4.105
55	0.279	0.912	4.127	3.894
60	0.359	1.064	4.448	3.743
65	0.460	1.276	4.747	3.531
70	0.573	1.460	5.035	3.346
75	0.700	1.548	5.308	3.259
80	0.831	1.588	5.578	3.219
85	0.961	1.582	5.849	3.224
90	1.093	1.635	6.117	3.172
95	1.233	1.683	6.378	3.124
100	1.374	1.711	6.637	3.095
105	1.518	1.731	6.894	3.075
110	1.663	1.755	7.150	3.051
115	1.810	1.770	7.404	3.037
120	1.958	1.795	7.656	3.011

SOIL TYPE - DOTHAN LOAMY SAND
IDENTIFICATION CODE - 04042W
COVER - BARE-80, WEEDS-20
DATE OF RUN - 11 05 69
RAINFALL INTENSITY - 6.370 INCHES/HOUR
INITIAL SOIL MOISTURE FOR THE 0 TO 12 INCH DEPTH - 2.78 INCHES
INITIAL SOIL MOISTURE FOR THE 12 TO 36 INCH DEPTH - 6.64 INCHES
FINAL SOIL MOISTURE FOR THE 0 TO 12 INCH DEPTH - 3.23 INCHES
FINAL SOIL MOISTURE FOR THE 12 TO 36 INCH DEPTH - 6.97 INCHES

TIME FROM START OF RAIN (MINUTES)	ACCUMULATED RUNOFF (INCHES)	RUNOFF RATE (IN/HR)	ACCUMULATED INFILTRATION (INCHES)	INFILTRATION RATE (IN/HR)
5	0.000	0.000	0.530	6.370
10	0.073	1.382	0.988	4.987
15	0.280	3.038	1.312	3.331
20	0.520	3.277	1.602	3.093
25	0.821	3.654	1.832	2.715
30	1.124	3.750	2.060	2.619
35	1.440	3.837	2.275	2.532
40	1.763	3.868	2.482	2.501
45	2.082	3.837	2.694	2.532
50	2.406	3.923	2.901	2.446
55	2.722	3.481	3.116	2.888
60	2.983	3.068	3.386	3.301
65	3.235	2.931	3.665	3.438
70	3.490	2.918	3.941	3.451
75	3.742	3.031	4.220	3.338
80	4.014	3.424	4.479	2.946
85	4.318	3.692	4.706	2.677
90	4.644	3.904	4.910	2.465
95	4.964	3.845	5.121	2.524
100	5.279	3.786	5.337	2.583
105	5.608	3.796	5.538	2.573
110	5.946	3.921	5.731	2.448
115	6.277	3.951	5.931	2.418
120	6.598	3.869	6.141	2.500

FUQUAY LO

Location: 0.3 mi west of Animal Disease Laboratory along hard surface road to junction with U.S. 41; northwest for 300 yd across pasture area and into cultivated field; Tift County, Ga.

Land use or cover: Corn.

Topography: Very gently sloping — 2½%.

Great soil group: Arenic plinthic paleudults; loamy, siliceous, thermic.

Parent material: Unconsolidated marine sediments of sandy clay loam.

Drainage: Well drained.

Horizon and Description

Ap: 0 to 10 inches. Dark-grayish-brown (10YR-4/2) loamy sand; weak, fine granular structure; very friable, nonsticky; numerous fine roots; strongly acid; abrupt smooth boundary.

A2: 7 to 28 inches. Light-yellowish-brown (2.5YR-6/4) loamy sand; weak, fine granular structure; very friable, nonsticky; fine roots common; very strongly acid; clear smooth boundary.

WEIGHT PERCENT AND V

DEPTH (inches)				TENSIONS (BARS)
	.1	.3	.6	3.
0-10	7.69	4.84	4.11	3.04
	12.23	7.70	6.53	4.83
	FRAGMENT	6.71		SIEVED
10-28	8.57	5.21	4.74	4.30
	12.60	7.66	6.97	6.32
	FRAGMENT	4.80		SIEVED
28-46	16.18	12.03	10.87	9.56
	26.86	19.97	18.04	15.87
	FRAGMENT	12.30		SIEVED
40-49	15.83	15.04	12.35	11.42
	25.17	23.91	19.64	18.16
	FRAGMENT	13.48		SIEVED
49+	15.78	12.67	12.43	11.38
	25.72	20.65	20.26	18.55
	FRAGMENT	0.00		SIEVED

1=FIST
2=CORE
3=LOOSE

SOIL TYPE - FUQUAY LOAMY SAND
IDENTIFICATION CODE - 05051D
COVER - BARE-50, WEEDS-50
DATE OF RUN - 10 19 69
RAINFALL INTENSITY - 4.687 INCHES/HOUR
INITIAL SOIL MOISTURE FOR THE 0 TO 12 INCH DEPTH - 0.82 INCHES
INITIAL SOIL MOISTURE FOR THE 12 TO 36 INCH DEPTH - 4.38 INCHES
FINAL SOIL MOISTURE FOR THE 0 TO 12 INCH DEPTH - 3.43 INCHES
FINAL SOIL MOISTURE FOR THE 12 TO 36 INCH DEPTH - 6.25 INCHES

TIME FROM START OF RAIN (MINUTES)	ACCUMULATED RUNOFF (INCHES)	RUNOFF RATE (IN/HR)	ACCUMULATED INFILTRATION (INCHES)	INFILTRATION RATE (IN/HR)
5	0.000	0.000	0.390	4.687
10	0.010	0.120	0.771	4.567
15	0.020	0.119	1.151	4.568
20	0.028	0.098	1.533	4.589
25	0.036	0.097	1.916	4.590
30	0.044	0.096	2.298	4.590
35	0.053	0.097	2.681	4.590
40	0.060	0.096	3.064	4.590
45	0.068	0.096	3.446	4.590
50	0.076	0.092	3.829	4.594
55	0.087	0.152	4.209	4.534
60	0.099	0.141	4.588	4.545
65	0.113	0.198	4.964	4.488
70	0.131	0.243	5.336	4.443
75	0.154	0.292	5.705	4.394
80	0.180	0.336	6.069	4.350
85	0.211	0.390	6.429	4.297
90	0.244	0.429	6.786	4.257
95	0.285	0.532	7.136	4.154
100	0.336	0.686	7.475	4.000
105	0.395	0.719	7.807	3.968
110	0.466	1.000	8.126	3.686
115	0.558	1.226	8.425	3.460
120	0.671	1.481	8.702	3.206
125	0.802	1.706	8.962	2.980
130	0.956	1.954	9.199	2.732
135	1.121	2.022	9.424	2.664
140	1.291	2.050	9.645	2.637
145	1.471	2.231	9.856	2.456
150	1.660	2.313	10.057	2.374
155	1.855	2.392	10.253	2.295
160	2.063	2.503	10.436	2.183
165	2.275	2.573	10.614	2.114
170	2.486	2.576	10.793	2.110
175	2.704	2.579	10.966	2.107
177	2.790	2.575	11.036	2.111

```
SOIL TYPE - FUQUAY LOAMY SAND
IDENTIFICATION CODE - 05051W
COVER - BARE-50,,WEEDS-50
DATE OF RUN - 10 19 69
RAINFALL INTENSITY - 3.245 INCHES/HOUR
INITIAL SOIL MOISTURE FOR THE 0 TO 12 INCH DEPTH - 2.84 INCHES
INITIAL SOIL MOISTURE FOR THE 12 TO 36 INCH DEPTH - 6.01 INCHES
FINAL SOIL MOISTURE FOR THE 0 TO 12 INCH DEPTH -  3.00 INCHES
FINAL SOIL MOISTURE FOR THE 12 TO 36 INCH DEPTH -  5.86 INCHES
```

TIME FROM START OF RAIN (MINUTES)	ACCUMULATED RUNOFF (INCHES)	RUNOFF RATE (IN/HR)	ACCUMULATED INFILTRATION (INCHES)	INFILTRATION RATE (IN/HR)
30	0.000	0.000	1.622	3.245
35	0.004	0.061	1.888	3.183
40	0.011	0.091	2.152	3.154
45	0.020	0.095	2.413	3.149
50	0.028	0.095	2.676	3.149
55	0.036	0.096	2.938	3.148
60	0.044	0.097	3.200	3.148
65	0.052	0.097	3.463	3.147
70	0.060	0.096	3.725	3.148
75	0.068	0.096	3.988	3.149
80	0.076	0.096	4.250	3.149
85	0.084	0.096	4.513	3.148
90	0.092	0.096	4.775	3.148
95	0.100	0.094	5.038	3.150
100	0.108	0.097	5.300	3.147
105	0.116	0.096	5.562	3.148
110	0.123	0.093	5.825	3.151
115	0.132	0.095	6.087	3.149
120	0.140	0.093	6.350	3.151
125	0.147	0.093	6.612	3.152
130	0.156	0.097	6.874	3.147
135	0.164	0.095	7.137	3.149
140	0.172	0.097	7.399	3.148
145	0.180	0.093	7.662	3.151

SOIL TYPE - FUQUAY LOAMY SAND
IDENTIFICATION CODE - 05052D
COVER - BARE-50, WEEDS-50
DATE OF RUN - 10 20 69
RAINFALL INTENSITY - 6.249 INCHES/HOUR
INITIAL SOIL MOISTURE FOR THE 0 TO 12 INCH DEPTH - 1.02 INCHES
INITIAL SOIL MOISTURE FOR THE 12 TO 36 INCH DEPTH - 3.64 INCHES
FINAL SOIL MOISTURE FOR THE 0 TO 12 INCH DEPTH - 3.47 INCHES
FINAL SOIL MOISTURE FOR THE 12 TO 36 INCH DEPTH - 6.37 INCHES

TIME FROM START OF RAIN (MINUTES)	ACCUMULATED RUNOFF (INCHES)	RUNOFF RATE (IN/HR)	ACCUMULATED INFILTRATION (INCHES)	INFILTRATION RATE (IN/HR)
4	0.000	0.000	0.416	6.249
5	0.004	0.480	0.516	5.769
10	0.080	0.429	0.961	5.820
15	0.116	0.323	1.446	5.926
20	0.140	0.250	1.943	5.999
25	0.160	0.237	2.444	6.011
30	0.180	0.217	2.944	6.032
35	0.196	0.187	3.449	6.061
40	0.212	0.189	3.954	6.060
45	0.228	0.189	4.459	6.060
50	0.244	0.189	4.963	6.060
55	0.260	0.181	5.469	6.068
60	0.276	0.211	5.973	6.038
65	0.295	0.234	6.474	6.015
70	0.316	0.239	6.974	6.010
75	0.340	0.441	7.471	5.808
80	0.392	0.751	7.941	5.498
85	0.464	0.960	8.389	5.289
90	0.552	1.182	8.821	5.067
95	0.667	1.578	9.228	4.670
100	0.813	2.004	9.602	4.245
105	0.998	2.414	9.939	3.835
110	1.214	2.819	10.244	3.430
115	1.461	3.027	10.516	3.222
120	1.718	3.087	10.781	3.162

SOIL TYPE - FUQUAY LOAMY SAND
IDENTIFICATION CODE - 05052W
COVER - WEEDS-50, BARE-50
DATE OF RUN - 10 20 69
RAINFALL INTENSITY - 5.168 INCHES/HOUR
INITIAL SOIL MOISTURE FOR THE 0 TO 12 INCH DEPTH - 2.71 INCHES
INITIAL SOIL MOISTURE FOR THE 12 TO 36 INCH DEPTH - 5.94 INCHES
FINAL SOIL MOISTURE FOR THE 0 TO 12 INCH DEPTH - 3.71 INCHES
FINAL SOIL MOISTURE FOR THE 12 TO 36 INCH DEPTH - 6.36 INCHES

TIME FROM START OF RAIN (MINUTES)	ACCUMULATED RUNOFF (INCHES)	RUNOFF RATE (IN/HR)	ACCUMULATED INFILTRATION (INCHES)	INFILTRATION RATE (IN/HR)
7	0.000	0.000	0.602	5.168
10	0.004	0.480	0.841	4.687
15	0.080	0.882	1.211	4.285
20	0.236	2.317	1.486	2.850
25	0.460	3.120	1.692	2.047
30	0.746	3.467	1.837	1.700
35	1.026	3.376	1.987	1.791
40	1.305	3.359	2.139	1.809
45	1.586	3.365	2.289	1.802
50	1.866	3.348	2.439	1.819
55	2.145	3.445	2.591	1.722
60	2.449	3.632	2.718	1.535
65	2.741	3.569	2.857	1.598
70	3.025	3.423	3.003	1.744
75	3.325	3.482	3.135	1.686
80	3.635	3.599	3.255	1.568
85	3.943	3.699	3.378	1.468
90	4.252	3.706	3.499	1.461
95	4.561	3.734	3.620	1.434
100	4.882	3.772	3.730	1.395
105	5.207	3.854	3.837	1.313

FUQUAY PEBBLY LOAMY SAND (06)

Location: 0.6 mi north of Oak Grove Church along U.S. 319; west along private road for 700 yd, 20 ft north of road; Tift County, Ga.
Land use or cover: Corn.
Topography: Very gently sloping—2½%.
Great soil group: Arenic plinthic paleudults; loamy, siliceous, thermic.
Parent material: Unconsolidated marine sediments of clay loam.
Drainage: Well drained.

Horizon and Description

Apcn: 0 to 9 inches. Dark grayish-brown (10YR–4/2) loamy sand; weak, fine granular structure; very friable, nonsticky; many small hard iron pebbles one-eighth to one-half inch in diameter; many fine roots; strongly acid; abrupt smooth boundary.

A2cn: 9 to 26 inches. Olive-yellow (2.5YR–6/6) loamy sand; weak, fine granular structure; very friable, nonsticky; many small hard iron pebbles; fine roots common; very strongly acid; clear smooth boundary.

B21tcn: 26 to 38 inches. Brownish-yellow (10YR–6/6) light sandy clay loam, weak, medium subangular blocky structure; friable, slightly sticky; many small hard iron pebbles; very strongly acid; gradual wavy boundary.

B22tcn: 38 to 48 inches. Light yellowish-brown (2.5YR–6/4 sandy clay loam with common medium distinct mottles of strong brown (7.5YR–5/6) and yellowish red (5YR–5/8); moderate, medium subangular blocky structure; friable, slightly sticky; many small hard iron pebbles; very strongly acid; gradual wavy boundary.

B23tpl: 48 to 60 inches. Reticulately mottled strong brown (7.5YR–5/8), light-gray (10YR–7/1), yellowish-red (5YR–5/8), and yellowish-brown (10YR–5/6) sandy clay loam; moderate, medium subangular blocky structure; firm, slightly sticky; few small hard iron pebbles; soft plinthite 10% to 30% by volume; very strongly acid.

Remarks: Colors are given for moist soil. Reaction determined by Soiltex.

FUQUAY PEBBLY LOAMY SAND (06)

WEIGHT PERCENT AND VOLUME PERCENT OF WATER RETAINED

DEPTH (inches)	TENSIONS (BARS)					BD G/CC	TP PCT	K
	.1	.3	.6	3.	15.			
0–9	7.39	4.79	3.73	3.21	2.89	1.90³	28.30	2.00–6.30
	14.04	9.10	7.09	6.10	5.49	0.00	0.00	
	FRAGMENT	0.00		SIEVED	3.07	ROCK PCT	16.30	
9–28	6.71	4.76	3.53	3.16	2.48	1.98³	25.28	2.00–6.30
	13.29	9.42	6.99	6.26	4.91	0.00	0.00	
	FRAGMENT	0.00		SIEVED	3.08	ROCK PCT	17.67	
28–40	9.21	8.08	7.53	7.07	6.53	1.68³	36.60	0.63–2.00
	15.47	13.57	12.65	11.88	10.97	0.00	0.00	
	FRAGMENT	0.00		SIEVED	6.22	ROCK PCT	39.02	
40–49	13.37	12.28	11.79	10.52	9.81	1.56³	41.13	0.63–2.00
	20.86	19.16	18.39	16.41	15.30	0.00	0.00	
	FRAGMENT	0.00		SIEVED	9.09	ROCK PCT	40.67	
49+	18.07	17.76	17.43	17.04	15.03	1.45³	45.28	0.63–2.00
	26.20	25.27	25.27	24.71	21.79	0.00	0.00	
	FRAGMENT	0.00		SIEVED	15.78	ROCK PCT	15.76	

1=FIST
2=CORE
3=LOOSE

SOIL TYPE - FUQUAY PEBBLY LOAMY SAND
IDENTIFICATION CODE - 06061D
COVER - GRASS-80, BARE-20
DATE OF RUN - 09 26 69
RAINFALL INTENSITY - 4.687 INCHES/HOUR
INITIAL SOIL MOISTURE FOR THE 0 TO 12 INCH DEPTH - 1.76 INCHES
INITIAL SOIL MOISTURE FOR THE 12 TO 36 INCH DEPTH - 5.22 INCHES
FINAL SOIL MOISTURE FOR THE 0 TO 12 INCH DEPTH - 2.52 INCHES
FINAL SOIL MOISTURE FOR THE 12 TO 36 INCH DEPTH - 5.92 INCHES

TIME FROM START OF RAIN (MINUTES)	ACCUMULATED RUNOFF (INCHES)	RUNOFF RATE (IN/HR)	ACCUMULATED INFILTRATION (INCHES)	INFILTRATION RATE (IN/HR)
3	0.000	0.000	0.234	4.687
5	0.020	1.201	0.350	3.485
10	0.145	1.637	0.635	3.049
15	0.272	1.503	0.899	3.184
20	0.400	1.492	1.162	3.195
25	0.520	1.429	1.432	3.258
30	0.642	1.455	1.701	3.231
35	0.761	1.426	1.972	3.260
40	0.882	1.590	2.242	3.096
45	1.022	1.640	2.492	3.046
50	1.144	1.367	2.761	3.319
55	1.249	1.037	3.047	3.650
60	1.320	0.773	3.366	3.913
65	1.391	0.936	3.686	3.751
70	1.475	0.981	3.993	3.705
75	1.549	0.913	4.309	3.774
80	1.630	0.960	4.619	3.726
85	1.717	1.244	4.923	3.442
90	1.823	1.033	5.207	3.654
95	1.883	0.739	5.537	3.947
100	1.952	0.783	5.859	3.904
105	2.009	0.762	6.193	3.924
110	2.084	0.892	6.509	3.794
115	2.155	0.975	6.828	3.712
120	2.242	0.572	7.132	4.115
125	2.266	0.695	7.498	3.991
130	2.374	1.272	7.781	3.415
135	2.473	1.077	8.073	3.609
140	2.561	1.084	8.375	3.602
145	2.645	1.095	8.682	3.592
150	2.737	1.070	8.980	3.617
155	2.811	0.913	9.297	3.773
160	2.889	0.908	9.610	3.778
164	2.950	0.921	9.861	3.765

SOIL TYPE — FUQUAY PEBBLY LOAMY SAND
IDENTIFICATION CODE — 06061W
COVER — GRASS-80, BARE-20
DATE OF RUN — 09 26 69
RAINFALL INTENSITY — 4.447 INCHES/HOUR
INITIAL SOIL MOISTURE FOR THE 0 TO 12 INCH DEPTH — 2.43 INCHES
INITIAL SOIL MOISTURE FOR THE 12 TO 36 INCH DEPTH — 5.83 INCHES
FINAL SOIL MOISTURE FOR THE 0 TO 12 INCH DEPTH — 2.45 INCHES
FINAL SOIL MOISTURE FOR THE 12 TO 36 INCH DEPTH — 6.05 INCHES

TIME FROM START OF RAIN (MINUTES)	ACCUMULATED RUNOFF (INCHES)	RUNOFF RATE (IN/HR)	ACCUMULATED INFILTRATION (INCHES)	INFILTRATION RATE (IN/HR)
3	0.000	0.000	0.222	4.447
5	0.012	1.322	0.334	3.124
10	0.113	1.188	0.628	3.258
15	0.196	0.920	0.915	3.526
20	0.267	0.736	1.215	3.710
25	0.331	0.817	1.521	3.629
30	0.401	0.815	1.822	3.631
35	0.468	0.809	2.125	3.637
40	0.540	0.883	2.424	3.563
45	0.613	0.878	2.722	3.568
50	0.687	0.943	3.017	3.503
55	0.766	0.984	3.310	3.462
60	0.848	1.027	3.598	3.419
65	0.936	1.129	3.881	3.317
70	1.031	1.197	4.156	3.249
75	1.139	1.358	4.419	3.088
80	1.260	1.556	4.668	2.890
85	1.394	1.633	4.905	2.813
90	1.531	1.624	5.138	2.822
92	1.585	1.625	5.232	2.821

SOIL TYPE - FUQUAY PEBBLY LOAMY SAND
IDENTIFICATION CODE - 06062D
COVER - GRASS-80, BARE-20
DATE OF RUN - 10 01 69
RAINFALL INTENSITY - 2.644 INCHES/HOUR
INITIAL SOIL MOISTURE FOR THE 0 TO 12 INCH DEPTH - 1.63 INCHES
INITIAL SOIL MOISTURE FOR THE 12 TO 36 INCH DEPTH - 4.96 INCHES
FINAL SOIL MOISTURE FOR THE 0 TO 12 INCH DEPTH - 2.82 INCHES
FINAL SOIL MOISTURE FOR THE 12 TO 36 INCH DEPTH - 6.08 INCHES

TIME FROM START OF RAIN (MINUTES)	ACCUMULATED RUNOFF (INCHES)	RUNOFF RATE (IN/HR)	ACCUMULATED INFILTRATION (INCHES)	INFILTRATION RATE (IN/HR)
12	0.000	0.000	0.528	2.644
15	0.004	0.300	0.645	2.343
20	0.044	0.361	0.837	2.283
25	0.073	0.347	1.027	2.297
30	0.115	0.406	1.206	2.238
35	0.149	0.408	1.392	2.236
40	0.181	0.376	1.581	2.268
45	0.214	0.383	1.769	2.260
50	0.246	0.384	1.957	2.260
55	0.277	0.379	2.146	2.264
60	0.310	0.385	2.334	2.258
65	0.341	0.376	2.523	2.267
70	0.374	0.380	2.709	2.264
75	0.403	0.332	2.901	2.311
80	0.431	0.337	3.094	2.306
85	0.459	0.337	3.286	2.306
90	0.487	0.344	3.478	2.299
95	0.514	0.332	3.671	2.311
100	0.543	0.337	3.863	2.307
105	0.571	0.337	4.055	2.306
110	0.599	0.340	4.247	2.303
115	0.627	0.342	4.439	2.301
120	0.656	0.353	4.631	2.290
125	0.684	0.349	4.823	2.295
130	0.713	0.355	5.015	2.288
135	0.741	0.362	5.207	2.281
140	0.768	0.348	5.400	2.296
145	0.796	0.340	5.593	2.303
150	0.824	0.348	5.785	2.296
155	0.852	0.340	5.978	2.303
160	0.879	0.328	6.171	2.315
163	0.896	0.336	6.286	2.308

SOIL TYPE — FUQUAY PEBBLY LOAMY SAND
IDENTIFICATION CODE — 06062W
COVER — GRASS-80, BARE-20
DATE OF RUN — 10 01 69
RAINFALL INTENSITY — 4.326 INCHES/HOUR
INITIAL SOIL MOISTURE FOR THE 0 TO 12 INCH DEPTH — 2.64 INCHES
INITIAL SOIL MOISTURE FOR THE 12 TO 36 INCH DEPTH — 5.92 INCHES
FINAL SOIL MOISTURE FOR THE 0 TO 12 INCH DEPTH — 2.72 INCHES
FINAL SOIL MOISTURE FOR THE 12 TO 36 INCH DEPTH — 6.05 INCHES

TIME FROM START OF RAIN (MINUTES)	ACCUMULATED RUNOFF (INCHES)	RUNOFF RATE (IN/HR)	ACCUMULATED INFILTRATION (INCHES)	INFILTRATION RATE (IN/HR)
4	0.000	0.000	0.288	4.326
5	0.016	0.961	0.344	3.365
10	0.096	0.972	0.625	3.354
15	0.170	0.838	0.911	3.488
20	0.239	0.862	1.202	3.464
25	0.311	0.857	1.491	3.469
30	0.383	0.864	1.779	3.462
35	0.455	0.862	2.068	3.464
40	0.528	0.866	2.356	3.459
45	0.601	0.883	2.643	3.443
50	0.672	0.873	2.932	3.453
55	0.746	0.929	3.219	3.397
60	0.827	1.029	3.499	3.297
65	0.916	1.124	3.771	3.202
70	1.010	1.155	4.037	3.171
75	1.106	1.146	4.301	3.180
80	1.203	1.158	4.565	3.168
85	1.299	1.145	4.830	3.181
89	1.377	1.160	5.040	3.166

SOIL TYPE - FUQUAY PEBBLY LOAMY SAND
IDENTIFICATION CODE - 06064D
COVER - GRASS-80, BARE-20
DATE OF RUN - 10 03 69
RAINFALL INTENSITY - 3.124 INCHES/HOUR
INITIAL SOIL MOISTURE FOR THE 0 TO 12 INCH DEPTH - 1.90 INCHES
INITIAL SOIL MOISTURE FOR THE 12 TO 36 INCH DEPTH - 5.64 INCHES
FINAL SOIL MOISTURE FOR THE 0 TO 12 INCH DEPTH - 2.76 INCHES
FINAL SOIL MOISTURE FOR THE 12 TO 36 INCH DEPTH - 5.75 INCHES

TIME FROM START OF RAIN (MINUTES)	ACCUMULATED RUNOFF (INCHES)	RUNOFF RATE (IN/HR)	ACCUMULATED INFILTRATION (INCHES)	INFILTRATION RATE (IN/HR)
9	0.000	0.000	0.468	3.124
10	0.012	0.721	0.508	2.403
15	0.068	0.792	0.713	2.332
20	0.136	0.996	0.905	2.128
25	0.216	0.960	1.085	2.164
30	0.296	0.958	1.266	2.166
35	0.376	0.960	1.446	2.164
40	0.457	0.975	1.626	2.149
45	0.537	0.918	1.806	2.206
50	0.610	0.855	1.993	2.269
55	0.678	0.805	2.186	2.319
60	0.741	0.771	2.383	2.353
65	0.805	0.764	2.580	2.360
70	0.869	0.771	2.776	2.353
75	0.934	0.781	2.971	2.343
80	0.997	0.766	3.169	2.358
85	1.061	0.764	3.365	2.360
90	1.125	0.762	3.562	2.362
95	1.190	0.777	3.757	2.347
100	1.253	0.767	3.954	2.357
105	1.317	0.759	4.150	2.365
110	1.381	0.758	4.347	2.366
115	1.447	0.783	4.542	2.341
120	1.510	0.765	4.739	2.359

SOIL TYPE – FUQUAY PEBBLY LOAMY SAND
IDENTIFICATION CODE – 06064W
COVER – GRASS-80, BARE-20
DATE OF RUN – 10 03 69
RAINFALL INTENSITY – 6.249 INCHES/HOUR
INITIAL SOIL MOISTURE FOR THE 0 TO 12 INCH DEPTH – 2.71 INCHES
INITIAL SOIL MOISTURE FOR THE 12 TO 36 INCH DEPTH – 5.78 INCHES
FINAL SOIL MOISTURE FOR THE 0 TO 12 INCH DEPTH – 2.84 INCHES
FINAL SOIL MOISTURE FOR THE 12 TO 36 INCH DEPTH – 6.82 INCHES

TIME FROM START OF RAIN (MINUTES)	ACCUMULATED RUNOFF (INCHES)	RUNOFF RATE (IN/HR)	ACCUMULATED INFILTRATION (INCHES)	INFILTRATION RATE (IN/HR)
2	0.000	0.000	0.208	6.249
5	0.052	3.605	0.340	2.644
10	0.524	3.910	0.516	2.338
15	0.822	3.591	0.740	2.658
20	1.132	3.767	0.950	2.482
25	1.441	3.744	1.162	2.505
30	1.757	3.800	1.367	2.449
35	2.064	3.707	1.581	2.542
40	2.386	3.830	1.780	2.419
45	2.688	3.715	1.998	2.534
50	3.000	3.683	2.208	2.565
55	3.317	3.719	2.411	2.530
60	3.630	3.749	2.619	2.500
65	3.944	3.753	2.826	2.496
70	4.248	3.668	3.043	2.581
75	4.559	3.650	3.252	2.599
80	4.875	3.695	3.457	2.554
85	5.190	3.771	3.663	2.478
90	5.486	3.279	3.888	2.970
95	5.778	3.483	4.117	2.766
100	6.088	3.798	4.327	2.451
105	6.408	3.793	4.528	2.456
110	6.702	3.906	4.755	2.343
115	7.056	3.884	4.922	2.365
120	7.390	3.846	5.109	2.403
125	7.692	3.785	5.327	2.464
130	8.001	3.759	5.539	2.490
135	8.314	3.700	5.747	2.548
140	8.645	3.954	5.937	2.295
145	8.952	3.866	6.151	2.383
150	9.255	3.764	6.369	2.485
155	9.600	4.109	6.544	2.140
160	9.884	3.807	6.782	2.442
165	10.215	4.001	6.971	2.248
170	10.524	4.032	7.183	2.217
175	10.847	4.093	'7.380	2.156

CONTINUED

CONTINUED

SOIL TYPE - FUQUAY PEBBLY LOAMY SAND
IDENTIFICATION CODE - 06064W

TIME FROM START OF RAIN (MINUTES)	ACCUMULATED RUNOFF (INCHES)	RUNOFF RATE (IN/HR)	ACCUMULATED INFILTRATION (INCHES)	INFILTRATION RATE (IN/HR)
180	11.169	4.210	7.579	2.039
185	11.479	4.091	7.791	2.157
190	11.775	3.831	8.015	2.418
195	12.131	4.209	8.180	2.040
200	12.436	4.066	8.396	2.183
205	12.738	3.798	8.615	2.451
210	13.074	3.995	8.799	2.254
215	13.382	3.849	9.012	2.400
220	13.724	4.095	9.191	2.154
225	14.017	3.795	9.419	2.454
230	14.325	3.624	9.632	2.625
235	14.655	3.746	9.823	2.503
240	14.984	3.833	10.014	2.415
245	15.289	3.632	10.230	2.617
250	15.611	3.677	10.428	2.572
255	15.939	3.780	10.622	2.469
260	16.264	3.852	10.818	2.397
265	16.578	3.772	11.024	2.477
270	16.913	3.880	11.210	2.369
275	17.222	3.795	11.422	2.454
280	17.556	3.935	11.609	2.314
285	17.845	3.552	11.840	2.697
290	18.147	3.365	12.060	2.884
295	18.494	3.650	12.233	2.599
300	18.825	3.776	12.423	2.472
305	19.133	3.603	12.636	2.646

KERSHAW COARSE SAND (07)

Location: 1.2 mi northwest of Whiddon Mill along county road; west along county road for 1.2 mi; north side of road; Tift County, Ga.
Land use or cover: Scrub live oak and turkey oak.
Topography: Gently sloping — 7%.
Great soil group: Typic quartzipsamments; thermic, uncoated.
Parent material: Unconsolidated beds of coarse sands.
Drainage: Very excessively drained.

Horizon and Description
A1: 0 to 3 inches. Dark grayish-brown (10YR-4/2) coarse sand with few medium, faint mottles of very dark grayish brown (10YR-3/2); structureless; loose; many fine and medium roots; very strongly acid; abrupt smooth boundary.

AC: 3 to 8 inches. Brown, dark-brown (10YR-4/3) coarse sand with common medium faint mottles of brownish yellow (10YR-6/6); structureless; loose; fine and medium roots common; very strongly acid; clear wavy boundary.

C1: 8 to 44 inches. Brownish-yellow (10YR-6/6) coarse sand; structureless; loose; few medium roots in upper part; very strongly acid; gradual wavy boundary.

C2: 44 to 80 inches. Pale-yellow (2.5YR-7/4) coarse sand with common coarse faint mottles of yellow (2.5YR-8/6); structureless; loose; very strongly acid.
Remarks: Colors are given for moist soil. Reaction determined by Soiltex.

KERSHAW COARSE SAND (07)

WEIGHT PERCENT AND VOLUME PERCENT OF WATER RETAINED

DEPTH (inches)	TENSIONS (BARS)					BD G/CC	TP PCT	K
	.1	.3	.6	3.	15.			
-3	6.51	5.12	4.69	3.08	1.69	1.79^3	32.45	6.30-20.00
	11.65	9.16	8.40	5.51	3.03	0.00	0.00	
	FRAGMENT	0.00		SIEVED	1.41	ROCK PCT	4.84	
-8	4.91	2.47	1.17	0.82	0.78	1.84^3	30.57	6.30-20.00
	9.03	4.54	2.15	1.51	1.44	0.00	0.00	
	FRAGMENT	0.00		SIEVED	0.80	ROCK PCT	0.87	
-44	2.60	1.48	1.04	0.96	0.46	2.12^3	20.00	6.30-20.00
	5.51	3.14	2.20	2.04	0.98	0.00	0.00	
	FRAGMENT	0.00		SIEVED	0.37	ROCK PCT	0.13	
44+	2.17	1.64	1.40	0.76	0.45	1.99^3	24.91	6.30-20.00
	4.32	3.26	2.79	1.51	0.90	0.00	0.00	
	FRAGMENT	0.00		SIEVED	0.39	ROCK PCT	0.07	

=FIST
=CORE
=LOOSE

SOIL TYPE — KERSHAW COARSE SAND
IDENTIFICATION CODE — 07071D
COVER — BARE-60, WEEDS-40
DATE OF RUN — 10 23 69
RAINFALL INTENSITY — 6.129 INCHES/HOUR
INITIAL SOIL MOISTURE FOR THE 0 TO 12 INCH DEPTH — 0.45 INCHES
INITIAL SOIL MOISTURE FOR THE 12 TO 36 INCH DEPTH — 1.25 INCHES
FINAL SOIL MOISTURE FOR THE 0 TO 12 INCH DEPTH — 3.10 INCHES
FINAL SOIL MOISTURE FOR THE 12 TO 36 INCH DEPTH — 5.25 INCHES

TIME FROM START OF RAIN (MINUTES)	ACCUMULATED RUNOFF (INCHES)	RUNOFF RATE (IN/HR)	ACCUMULATED INFILTRATION (INCHES)	INFILTRATION RATE (IN/HR)
7	0.000	0.000	0.715	6.129
10	0.016	0.841	0.977	5.288
15	0.100	0.752	1.432	5.377
20	0.152	0.469	1.890	5.660
25	0.180	0.348	2.373	5.781
30	0.212	0.327	2.852	5.802
35	0.232	0.139	3.343	5.989
40	0.240	0.164	3.845	5.965
45	0.260	0.201	4.336	5.927
50	0.272	0.139	4.835	5.990
55	0.284	0.122	5.334	6.006
60	0.292	0.068	5.837	6.061
65	0.296	0.041	6.344	6.088
70	0.300	0.051	6.850	6.078
75	0.303	0.042	7.358	6.087
80	0.308	0.049	7.864	6.080
85	0.312	0.018	8.371	6.110
90	0.312	0.022	8.882	6.106
95	0.316	0.019	9.389	6.109
100	0.316	0.027	9.899	6.102
105	0.320	0.027	10.406	6.102
110	0.320	0.025	10.917	6.104
115	0.324	0.025	11.423	6.104
120	0.324	0.028	11.934	6.101
125	0.328	0.023	12.441	6.106
130	0.328	0.023	12.952	6.106

SOIL TYPE - KERSHAW COARSE SAND
IDENTIFICATION CODE - 07071W
COVER - BARE-60, WEEDS-40
DATE OF RUN - 10 23 69
RAINFALL INTENSITY - 6.129 INCHES/HOUR
INITIAL SOIL MOISTURE FOR THE 0 TO 12 INCH DEPTH - 2.47 INCHES
INITIAL SOIL MOISTURE FOR THE 12 TO 36 INCH DEPTH - 4.60 INCHES
FINAL SOIL MOISTURE FOR THE 0 TO 12 INCH DEPTH - 3.00 INCHES
FINAL SOIL MOISTURE FOR THE 12 TO 36 INCH DEPTH - 5.22 INCHES

TIME FROM START OF RAIN (MINUTES)	ACCUMULATED RUNOFF (INCHES)	RUNOFF RATE (IN/HR)	ACCUMULATED INFILTRATION (INCHES)	INFILTRATION RATE (IN/HR)
12	0.000	0.000	1.225	6.129
15	0.008	0.360	1.520	5.769
20	0.040	0.362	2.003	5.766
25	0.072	0.385	2.481	5.743
30	0.104	0.396	2.960	5.733
35	0.136	0.339	3.439	5.790
40	0.160	0.258	3.926	5.871
45	0.180	0.236	4.416	5.893
50	0.200	0.237	4.907	5.891
55	0.220	0.242	5.398	5.886
60	0.240	0.204	5.889	5.925
65	0.256	0.236	6.384	5.893
70	0.280	0.279	6.871	5.849
75	0.300	0.163	7.361	5.965
80	0.308	0.082	7.864	6.047
85	0.316	0.074	8.367	6.055
90	0.320	0.044	8.874	6.085
95	0.324	0.048	9.380	6.080
100	0.328	0.049	9.887	6.080
105	0.332	0.050	10.394	6.079

SOIL TYPE - KERSHAW COARSE SAND
IDENTIFICATION CODE - 07072D
COVER - BARE-60, WEEDS-40
DATE OF RUN - 10 24 69
RAINFALL INTENSITY - 6.249 INCHES/HOUR
INITIAL SOIL MOISTURE FOR THE 0 TO 12 INCH DEPTH - 0.45 INCHES
INITIAL SOIL MOISTURE FOR THE 12 TO 36 INCH DEPTH - 1.48 INCHES
FINAL SOIL MOISTURE FOR THE 0 TO 12 INCH DEPTH - 2.69 INCHES
FINAL SOIL MOISTURE FOR THE 12 TO 36 INCH DEPTH - 5.43 INCHES

TIME FROM START OF RAIN (MINUTES)	ACCUMULATED RUNOFF (INCHES)	RUNOFF RATE (IN/HR)	ACCUMULATED INFILTRATION (INCHES)	INFILTRATION RATE (IN/HR)
4	0.000	0.000	0.416	6.249
5	0.001	0.084	0.519	6.165
10	0.008	0.078	1.033	6.171
15	0.012	0.071	1.550	6.177
20	0.020	0.099	2.063	6.150
25	0.028	0.095	2.576	6.154
30	0.036	0.096	3.088	6.153
35	0.044	0.097	3.601	6.152
40	0.052	0.097	4.114	6.152
45	0.060	0.096	4.627	6.152
50	0.068	0.096	5.140	6.153
55	0.076	0.100	5.652	6.149
60	0.084	0.072	6.165	6.177
65	0.088	0.044	6.682	6.205
70	0.092	0.048	7.199	6.201
75	0.096	0.047	7.716	6.202
80	0.100	0.046	8.233	6.203
85	0.104	0.051	8.749	6.198
90	0.108	0.046	9.266	6.203
95	0.112	0.048	9.783	6.201
100	0.116	0.045	10.300	6.204
105	0.120	0.050	10.816	6.199
110	0.124	0.033	11.334	6.216
115	0.126	0.021	11.852	6.228
120	0.128	0.025	12.371	6.224
125	0.130	0.023	12.890	6.226
130	0.132	0.021	13.409	6.228
135	0.134	0.022	13.928	6.227
140	0.136	0.026	14.446	6.223
145	0.138	0.026	14.965	6.223
150	0.140	0.022	15.484	6.227
155	0.141	0.021	16.003	6.228
160	0.144	0.022	16.522	6.227
165	0.146	0.026	17.040	6.223
170	0.148	0.017	17.559	6.232
175	0.149	0.016	18.079	6.233

CONTINUED

CONTINUED

SOIL TYPE — KERSHAW COARSE SAND
IDENTIFICATION CODE — 07072D

TIME FROM START OF RAIN (MINUTES)	ACCUMULATED RUNOFF (INCHES)	RUNOFF RATE (IN/HR)	ACCUMULATED INFILTRATION (INCHES)	INFILTRATION RATE (IN/HR)
180	0.150	0.016	18.598	6.233
185	0.152	0.013	19.118	6.236
190	0.153	0.019	19.637	6.230
195	0.155	0.017	20.156	6.232
200	0.156	0.015	20.676	6.234
205	0.157	0.015	21.196	6.234
210	0.159	0.018	21.715	6.231
215	0.160	0.012	22.235	6.237
220	0.161	0.019	22.754	6.230
225	0.162	0.014	23.274	6.235
230	0.164	0.015	23.793	6.234
235	0.165	0.014	24.313	6.234
240	0.166	0.014	24.832	6.235
245	0.167	0.010	25.352	6.239
250	0.169	0.019	25.871	6.230
255	0.171	0.017	26.390	6.232
260	0.172	0.016	26.910	6.233
265	0.173	0.012	27.430	6.237

LEEFIELD LOAMY SAND (08)

Location: 1.5 mi north of Coastal Plain Experiment Station dairy barn along station field roads; east for 425 yd along field road; south for 410 yd across cultivated field to within 50 ft of wooded area; Tift County, Ga.

Land use or cover: Corn.

Topography: Nearly level — 1½%.

Great soil group: Arenic plinthaquic paleudults; loamy, siliceous, thermic.

Parent material: Unconsolidated marine sediments of sandy clay loam.

Drainage: Somewhat poorly drained.

Horizon and Description

Ap: 0 to 10 inches. Dark-gray (10YR–4/1) loamy sand; weak, fine granular structure; very friable, non-sticky; many fine roots; strongly acid; abrupt smooth boundary.

A2g: 10 to 25 inches. Light brownish-gray (2.5YR–6/2) loamy sand with few fine faint mottles of light olive brown (2.5YR–5/6); weak, fine granular structure; very friable, nonsticky; few fine roots in upper part; very strongly acid; clear wavy boundary.

B21tg: 25 to 35 inches. Light yellowish-brown (2.5YR–6/4) sandy clay loam with common medium distinct mottles of light gray (10YR–7/1) and yellowish brown (10YR–5/8); moderate, medium subangular blocky structure; friable, slightly sticky; very strongly acid; clear wavy boundary.

B22tg: 35 to 48 inches. Light-gray (10YR–7/1) sandy clay loam with common coarse distinct mottles of yellowish brown (10YR–5/8) and strong brown (7.5YR–5/8); moderate, medium subangular blocky structure; friable, slightly sticky; few small hard iron pebbles; very strongly acid; clear wavy boundary.

B23tp1: 48 to 60 inches. Mottled light-gray (10YR–7/1), strong-brown (7.5YR–5/8), red (10YR–4/8), and yellowish-brown (10YR–5/8) sandy clay loam; moderate, medium subangular blocky structure; firm, slightly sticky; soft plinthite 10% to 20% by volume; very strongly acid.

Remarks: Colors are given for moist soil. Reaction determined by Soiltex.

LEEFIELD LOAMY SAND (08)

WEIGHT PERCENT AND VOLUME PERCENT OF WATER RETAINED

DEPTH (inches)	TENSIONS (BARS)					BD G/CC	TP PCT	%
	.1	.3	.6	3.	15.			
0-10	15.50	7.02	5.71	4.97	4.54	1.46[1]	44.91	2.00-6.30
	22.63	10.25	8.34	7.26	6.63	1.44	45.66	
	FRAGMENT	6.68		SIEVED	3.81	ROCK PCT	5.71	
10-25	13.03	7.29	5.28	4.62	4.14	1.56[1]	41.13	2.00-6.30
	20.33	11.37	8.24	7.21	6.46	1.57	40.75	
	FRAGMENT	7.58		SIEVED	3.85	ROCK PCT	8.48	
25-35	18.34	12.50	10.42	9.94	9.38	1.69[1]	36.23	0.63-2.00
	30.99	21.12	17.61	16.80	15.85	1.66	37.36	
	FRAGMENT	12.46		SIEVED	8.85	ROCK PCT	9.21	
35-48	20.32	18.59	10.88	8.85	8.68	1.70[1]	35.85	0.63-2.00
	34.54	31.60	18.50	15.04	14.76	1.71	35.47	
	FRAGMENT	18.06		SIEVED	6.93	ROCK PCT	12.94	
48+	16.49	12.39	11.94	10.00	9.06	1.68[1]	36.60	0.63-2.00
	27.70	20.82	20.06	16.80	15.22	1.64	38.11	
	FRAGMENT	11.62		SIEVED	8.86	ROCK PCT	7.47	

1=FIST
2=CORE
3=LOOSE

SOIL TYPE - LEEFIELD LOAMY SAND
IDENTIFICATION CODE - 08081D
COVER - WEEDS-80, BARE-20
DATE OF RUN - 10 29 69
RAINFALL INTENSITY - 6.500 INCHES/HOUR
INITIAL SOIL MOISTURE FOR THE 0 TO 12 INCH DEPTH - 1.08 INCHES
INITIAL SOIL MOISTURE FOR THE 12 TO 36 INCH DEPTH - 3.95 INCHES
FINAL SOIL MOISTURE FOR THE 0 TO 12 INCH DEPTH - 3.53 INCHES
FINAL SOIL MOISTURE FOR THE 12 TO 36 INCH DEPTH - 5.51 INCHES

TIME FROM START OF RAIN (MINUTES)	ACCUMULATED RUNOFF (INCHES)	RUNOFF RATE (IN/HR)	ACCUMULATED INFILTRATION (INCHES)	INFILTRATION RATE (IN/HR)
2	0.000	0.000	0.216	6.500
5	0.060	3.605	0.361	2.895
10	0.476	3.424	0.606	3.076
15	0.753	3.388	0.871	3.112
20	1.033	3.358	1.133	3.142
25	1.316	3.464	1.391	3.035
30	1.606	3.485	1.643	3.014
35	1.890	3.355	1.901	3.145
40	2.172	3.394	2.161	3.106
45	2.453	3.390	2.422	3.110
50	2.730	3.296	2.687	3.204
55	3.005	3.280	2.953	3.220
60	3.287	3.351	3.212	3.149
65	3.560	3.300	3.482	3.200
70	3.835	3.314	3.748	3.186
75	4.123	3.569	4.002	2.931
80	4.432	3.818	4.235	2.682
85	4.752	3.771	4.457	2.729
90	5.080	4.069	4.670	2.431
95	5.425	4.224	4.867	2.276
100	5.800	4.593	5.033	1.907
105	6.183	4.593	5.192	1.906
110	6.570	4.672	5.347	1.828
115	6.937	4.490	5.521	2.009
120	7.325	4.577	5.676	1.923

SOIL TYPE - LEEFIELD LOAMY SAND
IDENTIFICATION CODE - 08081W
COVER - WEEDS-80, BARE-20
DATE OF RUN - 10 29 69
RAINFALL INTENSITY - 5.168 INCHES/HOUR
INITIAL SOIL MOISTURE FOR THE 0 TO 12 INCH DEPTH - 2.87 INCHES
INITIAL SOIL MOISTURE FOR THE 12 TO 36 INCH DEPTH - 5.30 INCHES
FINAL SOIL MOISTURE FOR THE 0 TO 12 INCH DEPTH - 3.46 INCHES
FINAL SOIL MOISTURE FOR THE 12 TO 36 INCH DEPTH - 5.40 INCHES

TIME FROM START OF RAIN (MINUTES)	ACCUMULATED RUNOFF (INCHES)	RUNOFF RATE (IN/HR)	ACCUMULATED INFILTRATION (INCHES)	INFILTRATION RATE (IN/HR)
5	0.000	0.000	0.430	5.168
10	0.180	3.559	0.680	1.608
15	0.480	3.618	0.811	1.549
20	0.803	3.931	0.919	1.236
25	1.137	4.094	1.016	1.074
30	1.479	4.102	1.104	1.066
35	1.818	4.093	1.195	1.074
40	2.160	4.100	1.284	1.067
45	2.498	4.071	1.377	1.096
50	2.838	4.053	1.468	1.114
55	3.179	4.066	1.558	1.101
60	3.523	4.099	1.645	1.068
65	3.863	4.074	1.735	1.093
70	4.198	4.034	1.831	1.133
75	4.541	4.068	1.918	1.099
80	4.891	4.173	1.999	0.995
85	5.216	3.994	2.104	1.174
90	5.555	3.962	2.196	1.205
95	5.905	4.073	2.277	1.094
100	6.254	4.211	2.359	0.956

SOIL TYPE – LEEFIELD LOAMY SAND
IDENTIFICATION CODE – 08082D
COVER – WEEDS-80, BARE-20
DATE OF RUN – 10 30 69
RAINFALL INTENSITY – 4.807 INCHES/HOUR
INITIAL SOIL MOISTURE FOR THE 0 TO 12 INCH DEPTH – 1.30 INCHES
INITIAL SOIL MOISTURE FOR THE 12 TO 36 INCH DEPTH – 4.34 INCHES,
FINAL SOIL MOISTURE FOR THE 0 TO 12 INCH DEPTH – 3.09 INCHES
FINAL SOIL MOISTURE FOR THE 12 TO 36 INCH DEPTH – 5.67 INCHES

TIME FROM START OF RAIN (MINUTES)	ACCUMULATED RUNOFF (INCHES)	RUNOFF RATE (IN/HR)	ACCUMULATED INFILTRATION (INCHES)	INFILTRATION RATE (IN/HR)
9	0.000	0.000	0.721	4.807
10	0.008	0.600	0.793	4.206
15	0.088	1.092	1.113	3.714
20	0.192	1.266	1.410	3.541
25	0.291	1.080	1.711	3.727
30	0.371	0.855	2.032	3.952
35	0.440	0.783	2.364	4.024
40	0.508	0.875	2.696	3.932
45	0.589	1.020	3.016	3.786
50	0.678	1.137	3.327	3.670
55	0.774	1.127	3.632	3.680
60	0.863	1.097	3.944	3.710
65	0.953	1.060	4.255	3.747
70	1.040	1.143	4.568	3.663
75	1.150	1.472	4.859	3.334
80	1.277	1.354	5.133	3.453
85	1.377	1.312	5.432	3.495
90	1.503	1.517	5.708	3.290
95	1.626	1.507	5.985	3.300
100	1.757	1.616	6.254	3.191
105	1.885	1.544	6.527	3.262
110	2.015	1.548	6.798	3.258
115	2.144	1.587	7.069	3.219
120	2.282	1.756	7.332	3.050
125	2.431	1.885	7.584	2.922
130	2.597	2.035	7.819	2.772
135	2.758	2.000	8.058	2.807
140	2.925	2.002	8.291	2.804
145	3.095	2.012	8.522	2.795
150	3.265	1.997	8.753	2.810

SOIL TYPE - LEEFIELD LOAMY SAND
IDENTIFICATION CODE - 08082W
COVER - WEEDS-80, BARE-20
DATE OF RUN - 10 30 69
RAINFALL INTENSITY - 3.004 INCHES/HOUR
INITIAL SOIL MOISTURE FOR THE 0 TO 12 INCH DEPTH - 2.77 INCHES
INITIAL SOIL MOISTURE FOR THE 12 TO 36 INCH DEPTH - 5.52 INCHES
FINAL SOIL MOISTURE FOR THE 0 TO 12 INCH DEPTH - 2.95 INCHES
FINAL SOIL MOISTURE FOR THE 12 TO 36 INCH DEPTH - 5.63 INCHES

TIME FROM START OF RAIN (MINUTES)	ACCUMULATED RUNOFF (INCHES)	RUNOFF RATE (IN/HR)	ACCUMULATED INFILTRATION (INCHES)	INFILTRATION RATE (IN/HR)
14	0.000	0.000	0.701	3.004
15	0.000	0.060	0.750	2.944
20	0.006	0.060	0.995	2.944
25	0.010	0.056	1.241	2.947
30	0.016	0.075	1.486	2.929
35	0.024	0.117	1.728	2.886
40	0.036	0.164	1.967	2.840
45	0.052	0.244	2.201	2.759
50	0.076	0.294	2.427	2.710
55	0.100	0.308	2.654	2.696
60	0.128	0.364	2.876	2.640
65	0.160	0.410	3.094	2.594
70	0.196	0.457	3.309	2.547
75	0.236	0.480	3.519	2.524
80	0.275	0.470	3.730	2.534
85	0.315	0.474	3.940	2.530
90	0.355	0.475	4.151	2.528
95	0.396	0.471	4.361	2.533
100	0.436	0.499	4.571	2.505
105	0.481	0.544	4.776	2.460
110	0.525	0.536	4.982	2.468
115	0.569	0.553	5.189	2.451
120	0.618	0.620	5.391	2.384
125	0.670	0.643	5.589	2.360
130	0.723	0.649	5.786	2.355
135	0.773	0.627	5.987	2.377
140	0.825	0.621	6.185	2.383

ROBERTSDALE LOAMY SAND (09)

Location: 1 mi north of Coastal Plain Experiment Station dairy barn along station field roads; west for 0.3 mi along field road; 150 ft north of road in idle field; Tift County, Ga.

Land use or cover: Idle — gallberry, wiregrass, and common weeds.

Topography: Nearly level — 1%.

Great soil group: Plinthaquic fragiudults; fine-loamy siliceous, thermic.

Parent material: Unconsolidated marine sediments of sandy clay loam.

Drainage: Somewhat poorly drained.

Horizon and Description

Apcn: 0 to 6 inches. Dark-gray (10YR–4/1) loamy sand; weak, fine granular structure; very friable, nonsticky; common small hard iron pebbles one-eighth to one-half inch in diameter; many fine and medium roots; very strongly acid; abrupt smooth boundary.

A2cn: 6 to 25 inches. Light yellowish-brown (2.5YR–6/4) loamy sand with few fine distinct mottles of yellowish brown; weak, fine granular structure; very friable, nonsticky; small hard iron pebbles common; fine and medium roots common; very strongly acid; clear wavy boundary.

B1tgcn: 25 to 30 inches. Brownish-yellow (10YR–6/6) sandy loam with common medium distinct mottles of light yellowish brown (2.5YR–6/4) and light brownish gray (2.5YR–6/2); weak, medium granular structure; very friable, nonsticky; common small hard iron pebbles; few roots extend into layer; very strongly acid; clear wavy boundary.

B21tgcn: 30 to 44 inches. Mottled light yellowish-brown (2.5YR–6/4), light-gray (10YR–7/1), and red (10YR–4/8) sandy clay loam; many coarse mottles, distinct and prominent; massive structure; hard when dry, firm when moist, slightly sticky when wet; many small hard iron pebbles; very strongly acid; gradual wavy boundary.

B22tgcn: 44 to 60 inches. Yellowish-red (5YR–5/8) sandy clay loam with many coarse distinct mottles of light gray (10YR–7/1), red (10YR–4/8), and yellowish brown (10YR–5/8); massive structure; firm, slightly sticky; common hard iron pebbles; very strongly acid.

Remarks: Colors are given for moist soil. Reaction determined by Soiltex.

ROBERTSDALE LOAMY SAND (09)

WEIGHT PERCENT AND VOLUME PERCENT OF WATER RETAINED

DEPTH (inches)	TENSIONS (BARS)					BD G/CC	TP PCT	K
	.1	.3	.6	3.	15.			
0-6	13.11	9.24	8.62	4.76	3.76	1.39[1]	47.55	2.00-6.30
	18.22	12.84	11.98	6.62	5.23	1.49	43.77	
	FRAGMENT	8.84		SIEVED	3.84	ROCK PCT	6.16	
6-25	10.27	5.72	3.11	2.88	1.11	1.57[1]	40.75	2.00-6.30
	16.12	8.98	4.88	4.52	1.74	1.55	41.51	
	FRAGMENT	4.99		SIEVED	1.07	ROCK PCT	5.55	
25-30	12.58	9.11	8.18	7.85	5.33	1.58[1]	40.38	0.63-2.00
	19.88	14.39	12.92	12.40	8.42	1.58	40.38	
	FRAGMENT	7.17		SIEVED	4.76	ROCK PCT	23.58	
30-44	15.19	11.14	8.21	8.15	4.21	1.66[1]	37.36	0.63-2.00
	25.22	18.49	13.63	13.53	6.99	1.61	39.25	
	FRAGMENT	9.31		SIEVED	4.88	ROCK PCT	19.09	
44+	16.96	11.11	10.37	9.46	5.47	1.71[1]	35.47	0.63-2.00
	29.00	19.00	17.73	16.18	9.35	1.69	36.23	
	FRAGMENT	11.07		SIEVED	5.18	ROCK PCT	5.37	

1=FIST
2=CORE
3=LOOSE

SOIL TYPE - ROBERTSDALE LOAMY SAND
IDENTIFICATION CODE - 09091D
COVER - WEEDS-70, BARE-30
DATE OF RUN - 11 02 69
RAINFALL INTENSITY - 4.807 INCHES/HOUR
INITIAL SOIL MOISTURE FOR THE 0 TO 12 INCH DEPTH - 1.79 INCHES
INITIAL SOIL MOISTURE FOR THE 12 TO 36 INCH DEPTH - 5.51 INCHES
FINAL SOIL MOISTURE FOR THE 0 TO 12 INCH DEPTH - 2.54 INCHES
FINAL SOIL MOISTURE FOR THE 12 TO 36 INCH DEPTH - 6.40 INCHES

TIME FROM START OF RAIN (MINUTES)	ACCUMULATED RUNOFF (INCHES)	RUNOFF RATE (IN/HR)	ACCUMULATED INFILTRATION (INCHES)	INFILTRATION RATE (IN/HR)
4	0.000	0.000	0.320	4.807
5	0.024	1.442	0.376	3.365
10	0.172	2.517	0.629	2.289
15	0.400	2.884	0.801	1.922
20	0.641	2.925	0.961	1.881
25	0.891	3.127	1.111	1.680
30	1.159	3.289	1.244	1.518
35	1.435	3.426	1.369	1.380
40	1.728	3.543	1.477	1.264
45	2.018	3.537	1.587	1.270
50	2.319	3.587	1.686	1.220
55	2.612	3.554	1.794	1.253
60	2.907	3.541	1.899	1.265
65	3.201	3.485	2.006	1.321
70	3.504	3.554	2.104	1.252
75	3.795	3.504	2.213	1.303
80	4.093	3.564	2.316	1.243
85	4.394	3.580	2.416	1.226
90	4.698	3.671	2.513	1.136
95	5.022	3.818	2.589	0.989
100	5.327	3.782	2.684	1.025
105	5.641	3.752	2.771	1.055
110	5.973	3.897	2.840	0.910
115	6.290	3.846	2.924	0.961
120	6.618	3.854	2.996	0.953

SOIL TYPE - ROBERTSDALE LOAMY SAND
IDENTIFICATION CODE - 09091W
COVER - WEEDS-70, BARE-30
DATE OF RUN - 11 02 69
RAINFALL INTENSITY - 3.245 INCHES/HOUR
INITIAL SOIL MOISTURE FOR THE 0 TO 12 INCH DEPTH - 2.42 INCHES
INITIAL SOIL MOISTURE FOR THE 12 TO 36 INCH DEPTH - 6.26 INCHES
FINAL SOIL MOISTURE FOR THE 0 TO 12 INCH DEPTH - 2.54 INCHES
FINAL SOIL MOISTURE FOR THE 12 TO 36 INCH DEPTH - 6.44 INCHES

TIME FROM START OF RAIN (MINUTES)	ACCUMULATED RUNOFF (INCHES)	RUNOFF RATE (IN/HR)	ACCUMULATED INFILTRATION (INCHES)	INFILTRATION RATE (IN/HR)
4	0.000	0.000	0.216	3.245
5	0.008	1.081	0.262	2.163
10	0.152	2.132	0.388	1.112
15	0.344	2.221	0.466	1.023
20	0.537	2.371	0.544	0.873
25	0.737	2.441	0.614	0.804
30	0.941	2.459	0.680	0.785
35	1.145	2.448	0.747	0.796
40	1.351	2.465	0.812	0.779
45	1.553	2.446	0.880	0.798
50	1.761	2.487	0.943	0.758
55	1.963	2.470	1.010	0.775
60	2.168	2.456	1.076	0.789
65	2.371	2.443	1.143	0.802
70	2.574	2.406	1.211	0.839
75	2.773	2.376	1.283	0.868
80	2.976	2.373	1.350	0.871
85	3.179	2.422	1.417	0.822
90	3.379	2.388	1.488	0.857
95	3.571	2.288	1.566	0.956
100	3.770	2.309	1.638	0.936
105	3.968	2.332	1.710	0.913
110	4.161	2.273	1.787	0.972
115	4.360	2.325	1.859	0.919
120	4.559	2.358	1.930	0.886

SOIL TYPE — ROBERTSDALE LOAMY SAND
IDENTIFICATION CODE — 09092D
COVER — WEEDS-70, BARE-30
DATE OF RUN — 11 03 69
RAINFALL INTENSITY — 4.807 INCHES/HOUR
INITIAL SOIL MOISTURE FOR THE 0 TO 12 INCH DEPTH — 2.00 INCHES
INITIAL SOIL MOISTURE FOR THE 12 TO 36 INCH DEPTH — 5.67 INCHES
FINAL SOIL MOISTURE FOR THE 0 TO 12 INCH DEPTH — 2.68 INCHES
FINAL SOIL MOISTURE FOR THE 12 TO 36 INCH DEPTH — 5.93 INCHES

TIME FROM START OF RAIN (MINUTES)	ACCUMULATED RUNOFF (INCHES)	RUNOFF RATE (IN/HR)	ACCUMULATED INFILTRATION (INCHES)	INFILTRATION RATE (IN/HR)
2	0.000	0.000	0.160	4.807
5	0.020	2.524	0.296	2.283
10	0.328	2.872	0.473	1.935
15	0.557	2.799	0.644	2.007
20	0.791	2.880	0.811	1.927
25	1.035	2.969	0.967	1.837
30	1.282	3.092	1.121	1.715
35	1.548	3.285	1.256	1.522
40	1.829	3.349	1.375	1.458
45	2.108	3.389	1.497	1.418
50	2.408	3.600	1.598	1.207
55	2.702	3.587	1.704	1.220
60	3.000	3.537	1.806	1.270
65	3.303	3.577	1.905	1.230
70	3.602	3.569	2.006	1.238
75	3.909	3.612	2.099	1.195
80	4.201	3.556	2.208	1.250
85	4.496	3.469	2.314	1.338
90	4.808	3.595	2.403	1.211
95	5.105	3.568	2.506	1.238
100	5.403	3.537	2.609	1.270
105	5.694	3.506	2.719	1.301
110	6.022	3.799	2.791	1.007
115	6.324	3.679	2.889	1.127
120	6.632	3.690	2.982	1.117

SOIL TYPE - ROBERTSDALE LOAMY SAND
IDENTIFICATION CODE - 09092W
COVER - WEEDS-70, BARE-30
DATE OF RUN - 11 03 69
RAINFALL INTENSITY - 6.730 INCHES/HOUR
INITIAL SOIL MOISTURE FOR THE 0 TO 12 INCH DEPTH - 2.65 INCHES
INITIAL SOIL MOISTURE FOR THE 12 TO 36 INCH DEPTH - 6.32 INCHES
FINAL SOIL MOISTURE FOR THE 0 TO 12 INCH DEPTH - 2.86 INCHES
FINAL SOIL MOISTURE FOR THE 12 TO 36 INCH DEPTH - 6.98 INCHES

TIME FROM START OF RAIN (MINUTES)	ACCUMULATED RUNOFF (INCHES)	RUNOFF RATE (IN/HR)	ACCUMULATED INFILTRATION (INCHES)	INFILTRATION RATE (IN/HR)
2	0.000	0.000	0.224	6.730
5	0.036	4.086	0.384	2.644
10	0.565	5.213	0.556	1.517
15	1.001	5.302	0.680	1.428
20	1.442	5.359	0.801	1.370
25	1.895	5.478	0.908	1.252
30	2.352	5.463	1.012	1.266
35	2.799	5.402	1.126	1.327
40	3.235	5.256	1.252	1.473
45	3.679	5.221	1.368	1.509
50	4.121	5.230	1.487	1.500
55	4.567	5.247	1.602	1.483
60	5.003	5.251	1.726	1.478
65	5.449	5.320	1.841	1.410
70	5.891	5.296	1.960	1.434
75	6.326	5.316	2.087	1.414
80	6.792	5.495	2.181	1.234
85	7.242	5.590	2.291	1.140
90	7.698	5.607	2.396	1.122
95	8.164	5.639	2.491	1.090
100	8.616	5.551	2.601	1.179
105	9.090	5.720	2.687	1.010
110	9.522	5.400	2.816	1.329
115	9.999	5.584	2.900	1.146
120	10.483	5.843	2.977	0.887
125	10.938	5.799	3.083	0.930
130	11.385	5.589	3.197	1.141
135	11.840	5.519	3.303	1.210
140	12.316	5.708	3.388	1.021
145	12.778	5.714	3.487	1.016
150	13.214	5.443	3.611	1.287

STILSON LOAMY SAND (10)

Location: 1.5 mi north of Coastal Plain Experiment Station dairy barn along station road east for 425 yd along field road; south for 390 yd in cultivated field; Tift County, Ga.

Land use or cover: Corn.

Topography: Nearly level — 1%.

Great soil group: Arenic plinthic paleudults; loamy, siliceous, thermic.

Parent material: Unconsolidated marine sediments of sandy clay loam.

Drainage: Moderately well drained.

Horizon and Description

Ap: 0 to 9 inches. Dark-gray (10YR–4/1) loamy sand; weak fine granular structure; very friable, nonsticky; many fine roots; strongly acid; abrupt smooth boundary.

A2: 9 to 26 inches. Light yellowish-brown (2.5YR–6/4) loamy sand; weak, fine granular structure; very friable, nonsticky; fine roots common; strongly acid; clear wavy boundary.

B21t: 26 to 35 inches. Olive-yellow (2.5YR–6/6) sandy clay loam with common coarse distinct mottles of brownish yellow (10YR–6/6); weak, medium subangular blocky structure; friable, slightly sticky; very strongly acid; gradual wavy boundary.

B22tg: 35 to 45 inches. Light yellowish-brown (2.5YR–6/4) sandy clay loam with common medium distinc mottles of brownish yellow (10YR–6/6) and light gray (10YR–7/1); moderate, medium subangular blocky structure; friable; few small hard iron pebbles one eighth to one-half inch in diameter; very strongly acid; gradual wavy boundary.

B23tg: 45 to 53 inches. Light brownish-gray (2.5YR–6/2) sandy clay loam with many distinct and prominen coarse mottles of red (10YR–4/8) and strong brown (7.5YR–5/8); moderate, medium subangular blocky structure; firm, slightly sticky; few small hard iron pebbles; very strongly acid; gradual wavy boundary

B24tp1: 53 to 65 inches. Red (10YR–4/8) sandy clay loam with many coarse prominent mottles of light gray (10YR–7/1) and yellowish brown (10YR–5/8); moderate, medium subangular blocky structure; firm, slightly sticky; soft plinthite 10% to 20% by volume very strongly acid.

Remarks: Colors are given for moist soil. Reaction determined by Soiltex.

STILSON LOAMY SAND (10)

WEIGHT PERCENT AND VOLUME PERCENT OF WATER RETAINED

DEPTH (inches)	TENSIONS (BARS)					BD G/CC	TP PCT	u
	.1	.3	.6	3.	15.			
0–9	28.58	17.17	11.73	8.45	4.18	1.49[1]	43.77	2.00–6.30
	42.58	25.58	17.48	12.59	6.23	1.38	47.92	
	FRAGMENT	19.15		SIEVED	3.01	ROCK PERCENT	7.18	
9–26	12.86	11.06	10.67	6.84	3.89	1.47[1]	44.53	2.00–6.30
	18.90	16.26	15.68	10.05	5.72	1.48	44.15	
	FRAGMENT	11.52		SIEVED	2.47	ROCK PERCENT	5.46	
26–35	19.48	13.80	12.33	11.10	8.14	1.61[1]	39.25	0.63–2.00
	31.36	22.22	19.85	17.87	13.11	1.55	41.51	
	FRAGMENT	13.29		SIEVED	7.86	ROCK PERCENT	13.27	
35–45	20.64	14.82	13.67	11.91	11.40	1.59[1]	40.00	0.63–2.00
	32.82	23.56	21.74	18.94	18.13	1.59	40.00	
	FRAGMENT	12.96		SIEVED	10.23	ROCK PERCENT	12.64	
45–53	18.58	15.30	13.23	11.85	9.00	1.66[1]	37.36	0.63–2.00
	30.84	25.40	21.96	19.67	14.94	1.67	36.98	
	FRAGMENT	14.85		SIEVED	8.67	ROCK PERCENT	12.32	
53+	17.78	15.97	14.41	11.77	9.26	1.65[1]	37.74	0.63–2.00
	29.34	26.35	23.78	19.42	15.28	1.72	35.09	
	FRAGMENT	14.15		SIEVED	9.30	ROCK PERCENT	6.86	

1=FIST
2=CORE
3=LOOSE

SOIL TYPE - STILSON LOAMY SAND
IDENTIFICATION CODE - 10101D
COVER - WEEDS-50, BARE-50
DATE OF RUN - 10 27 69
RAINFALL INTENSITY - 6.249 INCHES/HOUR
INITIAL SOIL MOISTURE FOR THE 0 TO 12 INCH DEPTH - 1.22 INCHES
INITIAL SOIL MOISTURE FOR THE 12 TO 36 INCH DEPTH - 5.18 INCHES
FINAL SOIL MOISTURE FOR THE 0 TO 12 INCH DEPTH - 3.29 INCHES
FINAL SOIL MOISTURE FOR THE 12 TO 36 INCH DEPTH - 6.64 INCHES

TIME FROM START OF RAIN (MINUTES)	ACCUMULATED RUNOFF (INCHES)	RUNOFF RATE (IN/HR)	ACCUMULATED INFILTRATION (INCHES)	INFILTRATION RATE (IN/HR)
4	0.000	0.000	0.416	6.249
5	0.020	1.201	0.500	5.047
10	0.108	0.961	0.933	5.288
15	0.184	0.817	1.378	5.432
20	0.248	0.831	1.834	5.418
25	0.324	0.896	2.279	5.353
30	0.396	0.836	2.728	5.413
35	0.468	0.942	3.177	5.307
40	0.557	1.159	3.609	5.089
45	0.661	1.346	4.025	4.902
50	0.780	1.548	4.427	4.701
55	0.925	2.033	4.803	4.216
60	1.118	2.409	5.131	3.840
65	1.318	2.389	5.452	3.860
70	1.518	2.525	5.772	3.724
75	1.738	2.647	6.074	3.602
80	1.963	2.893	6.369	3.356
85	2.220	3.242	6.633	3.007
90	2.503	3.372	6.871	2.877
95	2.786	3.497	7.109	2.751
100	3.079	3.540	7.337	2.709
105	3.381	3.533	7.555	2.716
110	3.682	3.556	7.775	2.693
115	3.983	3.537	7.995	2.712
117	4.101	3.539	8.085	2.710

SOIL TYPE - STILSON LOAMY SAND
IDENTIFICATION CODE - 10101W
COVER - WEEDS-50, BARE-50
DATE OF RUN - 10 27 69
RAINFALL INTENSITY - 5.047 INCHES/HOUR
INITIAL SOIL MOISTURE FOR THE 0 TO 12 INCH DEPTH - 2.80 INCHES
INITIAL SOIL MOISTURE FOR THE 12 TO 36 INCH DEPTH - 6.38 INCHES
FINAL SOIL MOISTURE FOR THE 0 TO 12 INCH DEPTH - 3.26 INCHES
FINAL SOIL MOISTURE FOR THE 12 TO 36 INCH DEPTH - 6.58 INCHES

TIME FROM START OF RAIN (MINUTES)	ACCUMULATED RUNOFF (INCHES)	RUNOFF RATE (IN/HR)	ACCUMULATED INFILTRATION (INCHES)	INFILTRATIC RATE (IN/HR)
8	0.000	0.000	0.673	5.047
10	0.008	0.600	0.825	4.447
15	0.156	2.300	1.105	2.747
20	0.420	2.999	1.262	2.048
25	0.697	3.404	1.406	1.643
30	0.990	3.615	1.533	1.432
35	1.292	3.681	1.652	1.366
40	1.606	3.816	1.758	1.231
45	1.916	3.777	1.869	1.270
50	2.241	3.882	1.965	1.165
55	2.556	3.822	2.070	1.225
60	2.877	3.793	2.170	1.254
65	3.201	3.843	2.267	1.204
70	3.517	3.793	2.371	1.254
75	3.837	3.763	2.472	1.284
80	4.168	3.863	2.562	1.184
85	4.493	3.866	2.657	1.181
90	4.818	3.880	2.753	1.167
91	4.883	3.879	2.772	1.168

SOIL TYPE - STILSON LOAMY SAND
IDENTIFICATION CODE - 10102D
COVER - WEEDS-50, BARE-50
DATE OF RUN - 10 29 69
RAINFALL INTENSITY - 4.927 INCHES/HOUR
INITIAL SOIL MOISTURE FOR THE 0 TO 12 INCH DEPTH - 1.19 INCHES
INITIAL SOIL MOISTURE FOR THE 12 TO 36 INCH DEPTH - 4.48 INCHES
FINAL SOIL MOISTURE FOR THE 0 TO 12 INCH DEPTH - 3.41 INCHES
FINAL SOIL MOISTURE FOR THE 12 TO 36 INCH DEPTH - 6.18 INCHES

TIME FROM START OF RAIN (MINUTES)	ACCUMULATED RUNOFF (INCHES)	RUNOFF RATE (IN/HR)	ACCUMULATED INFILTRATION (INCHES)	INFILTRATION RATE (IN/HR)
6	0.000	0.000	0.492	4.927
10	0.008	0.120	0.813	4.807
15	0.018	0.119	1.213	4.807
20	0.028	0.228	1.614	4.699
25	0.048	0.244	2.005	4.683
30	0.068	0.240	2.395	4.687
35	0.088	0.239	2.786	4.688
40	0.108	0.242	3.176	4.685
45	0.128	0.241	3.567	4.686
50	0.148	0.241	3.958	4.686
55	0.168	0.239	4.348	4.687
60	0.188	0.239	4.739	4.688
65	0.207	0.234	5.130	4.693
70	0.228	0.232	5.520	4.695
75	0.248	0.278	5.911	4.649
80	0.276	0.381	6.293	4.546
85	0.311	0.481	6.669	4.446
90	0.359	0.723	7.031	4.204
95	0.433	0.889	7.368	4.037
100	0.505	0.919	7.707	4.008
105	0.584	0.951	8.038	3.976
110	0.665	1.057	8.368	3.870
115	0.760	1.206	8.683	3.720
120	0.867	1.334	8.987	3.593
125	0.979	1.373	9.286	3.554
130	1.089	1.377	9.587	3.549
135	1.209	1.483	9.878	3.443
140	1.339	1.575	10.158	3.352
145	1.470	1.612	10.437	3.315
150	1.607	1.667	10.711	3.260
155	1.746	1.707	10.983	3.220
160	1.890	1.722	11.250	3.205
165	2.035	1.751	11.515	3.176
170	2.183	1.771	11.777	3.156
175	2.333	1.785	12.039	3.142
180	2.477	1.757	12.305	3.170

SOIL TYPE — STILSON LOAMY SAND
IDENTIFICATION CODE — 10102W
COVER — WEEDS-50, BARE-50
DATE OF RUN — 10 29 69
RAINFALL INTENSITY — 3.004 INCHES/HOUR
INITIAL SOIL MOISTURE FOR THE 0 TO 12 INCH DEPTH — 2.96 INCHES
INITIAL SOIL MOISTURE FOR THE 12 TO 36 INCH DEPTH — 6.04 INCHES
FINAL SOIL MOISTURE FOR THE 0 TO 12 INCH DEPTH — 3.06 INCHES
FINAL SOIL MOISTURE FOR THE 12 TO 36 INCH DEPTH — 6.06 INCHES

TIME FROM START OF RAIN (MINUTES)	ACCUMULATED RUNOFF (INCHES)	RUNOFF RATE (IN/HR)	ACCUMULATED INFILTRATION (INCHES)	INFILTRATION RATE (IN/HR)
11	0.000	0.000	0.550	3.004
15	0.008	0.160	0.743	2.844
20	0.040	0.475	0.961	2.529
25	0.080	0.600	1.171	2.403
30	0.140	0.777	1.362	2.227
35	0.207	0.890	1.545	2.114
40	0.288	0.960	1.715	2.044
45	0.367	0.976	1.885	2.028
50	0.451	1.001	2.052	2.003
55	0.538	1.023	2.216	1.980
60	0.622	1.026	2.382	1.978
65	0.706	1.024	2.548	1.980
70	0.790	1.023	2.714	1.981
75	0.874	1.043	2.881	1.961
80	0.962	1.074	3.043	1.930
85	1.049	1.050	3.207	1.954
90	1.137	1.084	3.369	1.920
95	1.229	1.109	3.527	1.895
100	1.321	1.098	3.686	1.906
105	1.415	1.123	3.842	1.881
109	1.487	1.096	3.971	1.908

TROUP SAND (11)

Location: 0.5 mi east of Zion Hope Church along county road; north along county road for 0.2 mi; west along field road for 170 yd; 20 ft east of road in wooded area; Tift County, Ga.

Land use or cover: Pines.

Topography: Very gently sloping — 3%.

Great soil group: Grossarenic paleudults; loamy, siliceous, thermic.

Parent material: Unconsolidated marine sediments of sands and sandy clay loam.

Drainage: Excessively drained.

Horizon and Description

A1: 0 to 6 inches. Dark grayish-brown (10YR–4/2) sand; structureless; loose; many fine and medium roots; very strongly acid; abrupt smooth boundary.

A21: 6 to 36 inches. Yellowish-brown (10YR–5/4) sand; structureless; loose; fine and medium common mostly in upper part; very strongly acid; gradual wavy boundary.

A22: 36 to 55 inches. Light yellowish-brown (2.5YR–6/4) sand with few coarse faint mottles of pale yellow (2.5YR–8/4); structureless; loose; very strongly acid; clear wavy boundary.

Bt: 55 to 65 inches. Strong brown (7.5YR–5/8) sandy loam with few medium distinct mottles of yellowish brown (10YR–5/8) and yellowish red (5YR–4/8); weak medium granular structure; very friable; very strongly acid.

Remarks: Colors are given for moist soil. Reaction determined by Soiltex.

TROUP SAND (11)

WEIGHT PERCENT AND VOLUME PERCENT OF WATER RETAINED

DEPTH (inches)	TENSIONS (BARS)					BD G/CC	TP PCT	K
	.1	.3	.6	3.	15.			
0–6	11.24	4.99	2.42	1.59	0.69	1.72^3	35.09	6.30–20.00
	19.33	8.58	4.16	2.73	1.19	0.00	0.00	
	FRAGMENT	0.00		SIEVED	0.97	ROCK PCT	1.39	
6–36	11.67	4.58	1.61	1.36	0.59	1.75^3	33.96	6.30–20.00
	20.42	8.01	2.82	2.38	1.03	0.00	0.00	
	FRAGMENT	0.00		SIEVED	0.39	ROCK PCT	7.91	
36–55	10.30	2.88	1.78	1.57	0.27	1.73^3	34.72	6.30–20.00
	17.82	4.98	3.08	2.72	0.47	0.00	0.00	
	FRAGMENT	0.00		SIEVED	0.37	ROCK PCT	1.56	
55+	16.23	7.97	4.37	4.32	3.80	1.69^1	36.23	2.00–6.30
	27.43	13.47	7.39	7.30	6.42	1.64	38.11	
	FRAGMENT	8.78		SIEVED	3.93	ROCK PCT	4.44	

1=FIST
2=CORE
3=LOOSE

```
SOIL TYPE - TROUP SAND
IDENTIFICATION CODE - 11111D
COVER - GRASS-100
DATE OF RUN - 10 10 69
RAINFALL INTENSITY - 4.567 INCHES/HOUR
INITIAL SOIL MOISTURE FOR THE 0 TO 12 INCH DEPTH - 1.13 INCHES
INITIAL SOIL MOISTURE FOR THE 12 TO 36 INCH DEPTH - 5.27 INCHES
FINAL SOIL MOISTURE FOR THE 0 TO 12 INCH DEPTH -  3.05 INCHES
FINAL SOIL MOISTURE FOR THE 12 TO 36 INCH DEPTH -  6.59 INCHES
```

TIME FROM START OF RAIN (MINUTES)	ACCUMULATED RUNOFF (INCHES)	RUNOFF RATE (IN/HR)	ACCUMULATED INFILTRATION (INCHES)	INFILTRATION RATE (IN/HR)
4	0.000	0.000	0.304	4.567
5	0.032	1.802	0.348	2.764
10	0.236	2.439	0.524	2.128
15	0.408	2.022	0.733	2.544
20	0.577	1.985	0.944	2.581
25	0.737	1.871	1.166	2.696
30	0.891	1.823	1.392	2.744
35	1.039	1.773	1.624	2.793
40	1.182	1.714	1.862	2.852
45	1.322	1.629	2.103	2.937
50	1.454	1.560	2.351	3.006
55	1.583	1.510	2.603	3.056
60	1.706	1.487	2.860	3.080
65	1.828	1.453	3.119	3.114
70	1.951	1.423	3.377	3.143
75	2.073	1.402	3.635	3.165
80	2.193	1.430	3.896	3.136
85	2.322	1.552	4.147	3.015
90	2.446	1.554	4.404	3.013
95	2.593	1.879	4.637	2.687
100	2.763	2.232	4.848	2.334
105	2.958	2.428	5.034	2.138
110	3.158	2.390	5.215	2.176
115	3.369	2.444	5.383	2.122
120	3.589	2.547	5.544	2.020
125	3.813	2.631	5.701	1.936
130	4.044	2.674	5.851	1.893
134	4.220	2.659	5.979	1.907

SOIL TYPE - TROUP SAND
IDENTIFICATION CODE - 11111W
COVER - GRASS-100
DATE OF RUN - 10 10 69
RAINFALL INTENSITY - 6.500 INCHES/HOUR
INITIAL SOIL MOISTURE FOR THE 0 TO 12 INCH DEPTH - 2.73 INCHES
INITIAL SOIL MOISTURE FOR THE 12 TO 36 INCH DEPTH - 6.44 INCHES
FINAL SOIL MOISTURE FOR THE 0 TO 12 INCH DEPTH - 3.04 INCHES
FINAL SOIL MOISTURE FOR THE 12 TO 36 INCH DEPTH - 6.52 INCHES

TIME FROM START OF RAIN (MINUTES)	ACCUMULATED RUNOFF (INCHES)	RUNOFF RATE (IN/HR)	ACCUMULATED INFILTRATION (INCHES)	INFILTRATION RATE (IN/HR)
5	0.000	0.000	0.541	6.500
10	0.317	5.035	0.766	1.465
15	0.717	4.890	0.907	1.610
20	1.123	4.904	1.043	1.596
25	1.530	4.952	1.177	1.548
30	1.946	5.049	1.303	1.451
35	2.372	5.130	1.419	1.370
40	2.803	5.210	1.530	1.290
45	3.230	5.158	1.644	1.342
50	3.662	5.117	1.755	1.383
55	4.087	5.040	1.872	1.460
60	4.533	5.164	1.967	1.336
65	4.958	5.120	2.084	1.380
70	5.389	5.084	2.194	1.415
75	5.816	5.075	2.309	1.425
80	6.266	5.253	2.400	1.247
85	6.701	5.255	2.507	1.245
90	7.135	5.284	2.615	1.216
95	7.565	5.257	2.727	1.243
100	8.013	5.465	2.820	1.035
105	8.432	5.240	2.944	1.260
110	8.871	5.283	3.046	1.217
112	9.055	5.379	3.079	1.120

```
SOIL TYPE — TROUP SAND
IDENTIFICATION CODE — 11112D
COVER — GRASS-100
DATE OF RUN — 10 13 69
RAINFALL INTENSITY — 2.644 INCHES/HOUR
INITIAL SOIL MOISTURE FOR THE 0 TO 12 INCH DEPTH — 1.13 INCHES
INITIAL SOIL MOISTURE FOR THE 12 TO 36 INCH DEPTH — 4.91 INCHES
FINAL SOIL MOISTURE FOR THE 0 TO 12 INCH DEPTH — 2.63 INCHES
FINAL SOIL MOISTURE FOR THE 12 TO 36 INCH DEPTH — 6.11 INCHES
```

TIME FROM START OF RAIN (MINUTES)	ACCUMULATED RUNOFF (INCHES)	RUNOFF RATE (IN/HR)	ACCUMULATED INFILTRATION (INCHES)	INFILTRATION RATE (IN/HR)
5	0.000	0.000	0.220	2.644
10	0.050	1.207	0.390	1.436
15	0.128	1.125	0.532	1.519
20	0.216	1.021	0.664	1.622
25	0.300	0.976	0.801	1.668
30	0.376	0.805	0.945	1.838
35	0.436	0.715	1.106	1.928
40	0.497	0.653	1.265	1.990
45	0.544	0.568	1.438	2.075
50	0.593	0.561	1.609	2.082
55	0.638	0.525	1.785	2.118
60	0.678	0.464	1.966	2.179
65	0.714	0.442	2.150	2.202
70	0.749	0.439	2.335	2.204
75	0.786	0.421	2.518	2.222
80	0.817	0.381	2.708	2.262
85	0.850	0.397	2.895	2.246
90	0.882	0.371	3.083	2.272
95	0.909	0.332	3.277	2.312
100	0.936	0.328	3.470	2.316
105	0.965	0.316	3.661	2.328
110	0.991	0.303	3.856	2.340
115	1.013	0.282	4.054	2.361
120	1.038	0.305	4.249	2.338
125	1.061	0.262	4.447	2.381
130	1.081	0.227	4.647	2.416
135	1.100	0.227	4.848	2.417
140	1.121	0.233	5.048	2.411
145	1.141	0.237	5.248	2.406
150	1.161	0.232	5.449	2.411

SOIL TYPE — TROUP SAND
IDENTIFICATION CODE — 11112W
COVER — GRASS-100
DATE OF RUN — 10 13 69
RAINFALL INTENSITY — 4.567 INCHES/HOUR
INITIAL SOIL MOISTURE FOR THE 0 TO 12 INCH DEPTH — 2.36 INCHES
INITIAL SOIL MOISTURE FOR THE 12 TO 36 INCH DEPTH — 6.10 INCHES
FINAL SOIL MOISTURE FOR THE 0 TO 12 INCH DEPTH — 3.11 INCHES
FINAL SOIL MOISTURE FOR THE 12 TO 36 INCH DEPTH — 6.37 INCHES

TIME FROM START OF RAIN (MINUTES)	ACCUMULATED RUNOFF (INCHES)	RUNOFF RATE (IN/HR)	ACCUMULATED INFILTRATION (INCHES)	INFILTRATION RATE (IN/HR)
5	0.000	0.000	0.380	4.567
10	0.040	0.862	0.720	3.705
15	0.136	1.434	1.005	3.132
20	0.280	1.775	1.242	2.791
25	0.432	1.867	1.470	2.699
30	0.592	1.970	1.691	2.597
35	0.762	2.082	1.902	2.484
40	0.939	2.162	2.105	2.404
45	1.119	2.202	2.306	2.364
50	1.301	2.225	2.504	2.342
55	1.490	2.283	2.696	2.284
60	1.684	2.366	2.882	2.200
65	1.884	2.556	3.062	2.010
70	2.104	2.713	3.223	1.854
75	2.331	2.722	3.377	1.844
80	2.552	2.719	3.537	1.847
85	2.777	2.766	3.692	1.800
90	3.008	2.764	3.841	1.802
95	3.235	2.682	3.995	1.884
100	3.476	2.814	4.135	1.752
105	3.721	2.854	4.270	1.713
110	3.960	2.844	4.412	1.723
115	4.200	2.842	4.553	1.725
120	4.437	2.823	4.697	1.743
123	4.581	2.834	4.781	1.732

SOIL TYPE - TROUP SAND
IDENTIFICATION CODE - 11113D
COVER - GRASS-100
DATE OF RUN - 10 14 69
RAINFALL INTENSITY - 5.168 INCHES/HOUR
INITIAL SOIL MOISTURE FOR THE 0 TO 12 INCH DEPTH - 1.03 INCHES
INITIAL SOIL MOISTURE FOR THE 12 TO 36 INCH DEPTH - 3.88 INCHES
FINAL SOIL MOISTURE FOR THE 0 TO 12 INCH DEPTH - 2.85 INCHES
FINAL SOIL MOISTURE FOR THE 12 TO 36 INCH DEPTH - 6.19 INCHES

TIME FROM START OF RAIN (MINUTES)	ACCUMULATED RUNOFF (INCHES)	RUNOFF RATE (IN/HR)	ACCUMULATED INFILTRATION (INCHES)	INFILTRATION RATE (IN/HR)
3	0.000	0.000	0.258	5.168
5	0.016	1.201	0.394	3.966
10	0.136	1.209	0.724	3.959
15	0.236	1.193	1.055	3.974
20	0.365	1.337	1.357	3.830
25	0.478	1.356	1.674	3.812
30	0.591	1.333	1.992	3.835
35	0.698	1.239	2.316	3.928
40	0.795	1.137	2.649	4.030
45	0.885	1.055	2.990	4.112
50	0.968	0.996	3.338	4.171
55	1.054	1.024	3.683	4.143
60	1.137	1.017	4.030	4.150
65	1.220	0.995	4.378	4.173
70	1.304	0.991	4.725	4.176
75	1.390	1.020	5.069	4.147
80	1.474	1.026	5.416	4.141
85	1.558	1.014	5.763	4.153
90	1.641	1.011	6.110	4.156
95	1.723	0.970	6.459	4.197
100	1.808	0.981	6.805	4.186
105	1.891	0.975	7.152	4.193
110	1.976	0.978	7.498	4.189
115	2.062	0.989	7.843	4.178
120	2.144	0.968	8.192	4.199

SOIL TYPE - TROUP SAND
IDENTIFICATION CODE - 11113W
COVER - GRASS-100
DATE OF RUN - 10 14 69
RAINFALL INTENSITY - 2.764 INCHES/HOUR
INITIAL SOIL MOISTURE FOR THE 0 TO 12 INCH DEPTH - 2.31 INCHES
INITIAL SOIL MOISTURE FOR THE 12 TO 36 INCH DEPTH - 5.63 INCHES
FINAL SOIL MOISTURE FOR THE 0 TO 12 INCH DEPTH - 2.61 INCHES
FINAL SOIL MOISTURE FOR THE 12 TO 36 INCH DEPTH - 5.89 INCHES

TIME FROM START OF RAIN (MINUTES)	ACCUMULATED RUNOFF (INCHES)	RUNOFF RATE (IN/HR)	ACCUMULATED INFILTRATION (INCHES)	INFILTRATION RATE (IN/HR)
17	0.000	0.000	0.783	2.764
20	0.002	0.120	0.915	2.644
25	0.016	0.120	1.135	2.644
30	0.025	0.177	1.356	2.586
35	0.044	0.284	1.567	2.479
40	0.071	0.338	1.771	2.426
45	0.099	0.336	1.973	2.427
50	0.127	0.336	2.176	2.427
55	0.155	0.338	2.378	2.426
60	0.183	0.332	2.581	2.431
65	0.211	0.332	2.783	2.432
70	0.240	0.360	2.984	2.403
75	0.270	0.353	3.185	2.410
80	0.301	0.386	3.384	2.377
85	0.332	0.372	3.583	2.392
90	0.366	0.421	3.779	2.342
95	0.400	0.407	3.976	2.356
100	0.435	0.439	4.171	2.325
105	0.471	0.429	4.366	2.334
110	0.507	0.433	4.560	2.331
115	0.543	0.429	4.755	2.335
120	0.579	0.438	4.948	2.325

TROUP SAND (12)

Location: 0.6 mi north of Oak Grove Church along U.S. 319; west along private road for 800 yd; 20 ft north of road; Tift County, Ga.

Land use or cover: Corn.

Topography: Very gently sloping — 2%.

Great soil group: Grossarenic paleudults; loamy, siliceous, thermic.

Parent material: Unconsolidated marine sediments of sands and sandy clay loam.

Drainage: Excessively drained.

Horizon and Description

Ap: 0 to 10 inches. Dark grayish-brown (10YR–4/2) sand; structureless; loose; many fine roots; very strongly acid; abrupt smooth boundary.

A21: 10 to 42 inches. Light yellowish-brown (2.5YR–6/4) sand; structureless; loose; fine roots common in upper part mostly; very strongly acid; gradual wavy boundary.

A22: 42 to 53 inches. Light yellowish-brown (2.5YR–6/4) sand with common coarse distinct mottles of yellowish brown (10YR–5/8) and pale yellow (2.5YR–7/4); structureless; loose; very strongly acid; clear wavy boundary.

B2t: 53 to 65 inches. Yellowish-brown (10YR–5/8) sandy clay loam with common medium-distinct and prominent mottles of yellowish red (5YR–4/8), light gray (10YR–7/1), and red (2.5YR–4/8); moderate, medium subangular blocky structure; firm, slightly sticky; very strongly acid.

Remarks: Colors are given for moist soil. Reaction determined by Soiltex.

WEIGHT PERCENT AND VOLUME PERCENT OF WATER RETAINED

DEPTH (inches)	TENSIONS (BARS)					BD G/CC	TP PCT	K
	.1	.3	.6	3.	15.			
0–10	6.88	1.84	1.47	0.66	0.40	1.69[2]	36.23	6.30–20.00
	11.63	3.11	2.48	1.12	0.68	0.00	0.00	
	FRAGMENT	0.00		SIEVED	0.34	ROCK PCT	1.31	
10–42	6.53	2.36	1.56	0.61	0.40	1.65[2]	37.74	6.30–20.00
	10.77	3.89	2.57	1.01	0.66	0.00	0.00	
	FRAGMENT	0.00		SIEVED	0.44	ROCK PCT	1.56	
42–53	7.02	2.01	0.99	0.45	0.36	1.65[2]	37.74	6.30–20.00
	11.58	3.32	1.63	0.74	0.59	0.00	0.00	
	FRAGMENT	0.00		SIEVED	0.39	ROCK PCT	1.77	
53+	18.93	12.82	11.76	10.99	9.02	1.66[1]	37.36	0.63–2.00
	31.42	21.28	19.52	18.24	14.97	1.63	38.49	
	FRAGMENT	11.10		SIEVED	8.62	ROCK PCT	2.77	

1=FIST
2=CORE
3=LOOSE

SOIL TYPE - TROUP SAND
IDENTIFICATION CODE - 12112D
COVER - WEEDS-60, BARE-40
DATE OF RUN - 09 23 69
RAINFALL INTENSITY - 6.249 INCHES/HOUR
INITIAL SOIL MOISTURE FOR THE 0 TO 12 INCH DEPTH - 1.73 INCHES
INITIAL SOIL MOISTURE FOR THE 12 TO 36 INCH DEPTH - 3.39 INCHES
FINAL SOIL MOISTURE FOR THE 0 TO 12 INCH DEPTH - 3.56 INCHES
FINAL SOIL MOISTURE FOR THE 12 TO 36 INCH DEPTH - 6.00 INCHES

TIME FROM START OF RAIN (MINUTES)	ACCUMULATED RUNOFF (INCHES)	RUNOFF RATE (IN/HR)	ACCUMULATED INFILTRATION (INCHES)	INFILTRATION RATE (IN/HR)
3	0.000	0.000	0.312	6.249
5	0.004	3.725	0.448	2.524
10	0.365	3.711	0.675	2.538
15	0.683	3.734	0.878	2.515
20	0.991	3.710	1.091	2.539
25	1.298	3.651	1.305	2.598
30	1.603	3.668	1.521	2.581
35	1.906	3.616	1.739	2.633
40	2.208	3.604	1.957	2.645
45	2.503	3.491	2.184	2.758
50	2.788	3.441	2.419	2.808
55	3.076	3.426	2.652	2.823
60	3.366	3.451	2.882	2.797
65	3.656	3.464	3.114	2.785
70	3.946	3.463	3.345	2.786
75	4.228	3.424	3.583	2.825
80	4.518	3.436	3.814	2.813
85	4.810	3.478	4.043	2.771
90	5.096	3.455	4.277	2.794
95	5.384	3.407	4.511	2.842
100	5.673	3.448	4.742	2.801
105	5.965	3.484	4.971	2.764
110	6.250	3.454	5.207	2.795
115	6.538	3.434	5.440	2.815
120	6.833	3.543	5.665	2.706
125	7.116	3.456	5.904	2.792
130	7.399	3.393	6.141	2.856
135	7.695	3.493	6.366	2.756
140	7.982	3.465	6.600	2.784
145	8.274	3.533	6.828	2.716
150	8.559	3.470	7.064	2.779

SOIL TYPE — TROUP SAND
IDENTIFICATION CODE — 12112W
COVER — WEEDS-60, BARE-40
DATE OF RUN — 09 23 69
RAINFALL INTENSITY — 6.249 INCHES/HOUR
INITIAL SOIL MOISTURE FOR THE 0 TO 12 INCH DEPTH — 3.10 INCHES
INITIAL SOIL MOISTURE FOR THE 12 TO 36 INCH DEPTH — 5.69 INCHES
FINAL SOIL MOISTURE FOR THE 0 TO 12 INCH DEPTH — 3.67 INCHES
FINAL SOIL MOISTURE FOR THE 12 TO 36 INCH DEPTH — 6.86 INCHES

TIME FROM START OF RAIN (MINUTES)	ACCUMULATED RUNOFF (INCHES)	RUNOFF RATE (IN/HR)	ACCUMULATED INFILTRATION (INCHES)	INFILTRATION RATE (IN/HR)
6	0.000	0.000	0.625	6.249
10	0.288	4.559	0.753	1.690
15	0.685	4.596	0.877	1.653
20	1.062	4.370	1.021	1.879
25	1.415	4.361	1.188	1.888
30	1.789	4.525	1.335	1.724
35	2.167	4.515	1.478	1.734
40	2.543	4.498	1.622	1.751
45	2.916	4.494	1.771	1.755
50	3.294	4.464	1.914	1.785
55	3.672	4.486	2.056	1.763
60	4.052	4.493	2.197	1.756
65	4.424	4.464	2.345	1.785
70	4.795	4.404	2.495	1.845
75	5.178	4.505	2.634	1.744
80	5.545	4.406	2.787	1.843
85	5.934	4.529	2.919	1.720
90	6.317	4.619	3.056	1.630

SOIL TYPE - TROUP SAND
IDENTIFICATION CODE - 12113D
COVER - WEEDS-60, BARE-40
DATE OF RUN - 09 24 69
RAINFALL INTENSITY - 6.500 INCHES/HOUR
INITIAL SOIL MOISTURE FOR THE 0 TO 12 INCH DEPTH - 2.03 INCHES
INITIAL SOIL MOISTURE FOR THE 12 TO 36 INCH DEPTH - 5.59 INCHES
FINAL SOIL MOISTURE FOR THE 0 TO 12 INCH DEPTH - 3.20 INCHES
FINAL SOIL MOISTURE FOR THE 12 TO 36 INCH DEPTH - 7.12 INCHES

TIME FROM START OF RAIN (MINUTES)	ACCUMULATED RUNOFF (INCHES)	RUNOFF RATE (IN/HR)	ACCUMULATED INFILTRATION (INCHES)	INFILTRATION RATE (IN/HR)
6	0.000	0.000	0.650	6.500
10	0.040	0.601	1.043	5.898
15	0.090	0.598	1.535	5.901
20	0.154	0.828	2.012	5.672
25	0.225	1.231	2.483	5.269
30	0.394	2.506	2.856	3.994
35	0.612	2.864	3.179	3.636
40	0.881	3.550	3.452	2.950
45	1.196	3.947	3.679	2.553
50	1.532	4.159	3.884	2.341
55	1.892	4.393	4.066	2.107
60	2.254	4.334	4.246	2.165
65	2.609	4.295	4.433	2.204
70	2.971	4.294	4.613	2.205
75	3.327	4.249	4.798	2.251
80	3.693	4.296	4.974	2.204
85	4.048	4.254	5.160	2.246
90	4.412	4.258	5.339	2.242
95	4.779	4.368	5.513	2.131
100	5.131	4.279	5.703	2.221
105	5.490	4.233	5.886	2.267
110	5.841	4.138	6.076	2.361
115	6.225	4.414	6.234	2.085
120	6.582	4.358	6.419	2.142
125	6.951	4.481	6.591	2.019
130	7.300	4.335	6.784	2.165
135	7.663	4.366	6.963	2.134
140	8.023	4.374	7.144	2.126
145	8.398	4.526	7.311	1.973
150	8.740	4.318	7.511	2.182

SOIL TYPE - TROUP SAND
IDENTIFICATION CODE - 12113W
COVER - WEEDS-60, BARE-40
DATE OF RUN - 09 24 69
RAINFALL INTENSITY - 2.884 INCHES/HOUR
INITIAL SOIL MOISTURE FOR THE 0 TO 12 INCH DEPTH - 2.87 INCHES
INITIAL SOIL MOISTURE FOR THE 12 TO 36 INCH DEPTH - 6.73 INCHES
FINAL SOIL MOISTURE FOR THE 0 TO 12 INCH DEPTH - 2.92 INCHES
FINAL SOIL MOISTURE FOR THE 12 TO 36 INCH DEPTH - 6.83 INCHES

TIME FROM START OF RAIN (MINUTES)	ACCUMULATED RUNOFF (INCHES)	RUNOFF RATE (IN/HR)	ACCUMULATED INFILTRATION (INCHES)	INFILTRATION RATE (IN/HR)
4	0.000	0.000	0.192	2.884
5	0.020	1.201	0.220	1.682
10	0.120	1.665	0.360	1.219
15	0.264	1.684	0.456	1.200
20	0.405	1.759	0.555	1.125
25	0.555	1.792	0.646	1.092
30	0.703	1.793	0.738	1.091
35	0.851	1.783	0.831	1.101
40	0.999	1.788	0.923	1.096
45	1.147	1.744	1.016	1.139
50	1.290	1.713	1.113	1.171
55	1.428	1.559	1.215	1.324
60	1.556	1.568	1.328	1.316
65	1.688	1.584	1.436	1.300
70	1.819	1.563	1.546	1.321
75	1.953	1.592	1.652	1.291
80	2.084	1.578	1.761	1.306
85	2.219	1.610	1.866	1.273
88	2.300	1.618	1.930	1.266

- TROUP SAND
ATION CODE - 12114D
EEDS-60, BARE-40
JN - 09 25 69
INTENSITY - 3.846 INCHES/HOUR
JIL MOISTURE FOR THE 0 TO 12 INCH DEPTH - 2.65 INCHES
JIL MOISTURE FOR THE 12 TO 36 INCH DEPTH - 8.57 INCHES
L MOISTURE FOR THE 0 TO 12 INCH DEPTH - 3.87 INCHES
L MOISTURE FOR THE 12 TO 36 INCH DEPTH - 11.20 INCHES

JM RAIN S)	ACCUMULATED RUNOFF (INCHES)	RUNOFF RATE (IN/HR)	ACCUMULATED INFILTRATION (INCHES)	INFILTRATION RATE (IN/HR)
	0.000	0.000	0.256	3.846
	0.020	0.961	0.300	2.884
	0.132	1.484	0.508	2.361
	0.264	1.624	0.697	2.221
	0.400	1.633	0.881	2.212
	0.537	1.637	1.065	2.208
	0.675	1.662	1.247	2.183
	0.810	1.650	1.433	2.195
	0.945	1.633	1.618	2.212
	1.082	1.642	1.802	2.203
	1.218	1.641	1.986	2.204
	1.353	1.621	2.172	2.224
	1.491	1.644	2.354	2.202
	1.626	1.654	2.539	2.192
	1.768	1.729	2.719	2.116
	1.907	1.721	2.899	2.124
	2.057	1.816	3.070	2.030
	2.215	1.919	3.232	1.926
	2.378	1.955	3.390	1.890
	2.532	1.869	3.557	1.976
	2.694	1.923	3.715	1.922
	2.858	1.938	3.872	1.908
	3.022	1.956	4.028	1.889
	3.186	1.989	4.184	1.857
	3.359	1.972	4.332	1.873
	3.530	1.965	4.482	1.881
	3.708	2.020	4.624	1.825
	3.885	2.077	4.768	1.768
	4.053	2.032	4.920	1.813
	4.222	2.007	5.072	1.838
	4.399	2.085	5.215	1.760
	4.570	2.079	5.364	1.766
	4.751	2.169	5.504	1.676
	4.913	2.038	5.662	1.807
	5.092	2.125	5.804	1.720
	5.266	2.159	5.951	1.686
	5.428	1.996	6.109	1.849
	5.609	2.004	6.249	1.841
	5.788	2.090	6.390	1.755
	5.963	2.043	6.535	1.802
	6.148	2.132	6.671	1.713
	6.294	2.188	6.782	1.657

SOIL TYPE - TROUP SAND
IDENTIFICATION CODE - 12114W
COVER - WEEDS-60, BARE-40
DATE OF RUN - 09 25 69
RAINFALL INTENSITY - 3.846 INCHES/HOUR
INITIAL SOIL MOISTURE FOR THE 0 TO 12 INCH DEPTH - 3.38 INCHES
INITIAL SOIL MOISTURE FOR THE 12 TO 36 INCH DEPTH - 9.80 INCHES
FINAL SOIL MOISTURE FOR THE 0 TO 12 INCH DEPTH - 3.65 INCHES
FINAL SOIL MOISTURE FOR THE 12 TO 36 INCH DEPTH - 10.93 INCHES

TIME FROM START OF RAIN (MINUTES)	ACCUMULATED RUNOFF (INCHES)	RUNOFF RATE (IN/HR)	ACCUMULATED INFILTRATION (INCHES)	INFILTRATION RATE (IN/HR)
2	0.000	0.000	0.128	3.846
5	0.032	2.403	0.220	1.442
10	0.320	2.566	0.320	1.279
15	0.533	2.673	0.427	1.172
20	0.749	2.657	0.532	1.189
25	0.972	2.734	0.629	1.111
30	1.203	2.799	0.719	1.046
35	1.435	2.795	0.807	1.050
40	1.673	2.884	0.890	0.961
45	1.907	2.833	0.977	1.012
50	2.143	2.818	1.061	1.027
55	2.382	2.868	1.143	0.977
60	2.616	2.837	1.229	1.008

TIFTON LOAMY SAND (13)

Location: 350 yd northwest of superintendent's house on Coastal Plain Experiment Station Agronomy Farm along field road; 15 ft north of road in cultivated field; Tift County, Ga.
Land use or cover: Cotton.
Topography: Nearly level — 1%.
Great soil group: Plinthic paleudults; fine-loamy, siliceous, thermic.
Parent material: Unconsolidated marine sediments of sandy clay loam.
Drainage: Well drained.

Horizon and Description

Apcn: 0 to 11 inches. Very dark grayish-brown (2.5YR-3/2) loamy sand; weak, fine granular structure; very friable, nonsticky; many small hard iron pebbles one-eighth to one-half inch in diameter; many fine roots; strongly acid; abrupt smooth boundary.

B1tcn: 11 to 17 inches. Yellowish-brown (10YR-5/8) sandy clay loam; weak, medium subangular blocky structure; friable, slightly sticky; many small hard iron pebbles; fine roots common; strongly acid; clear wavy boundary.

B23tcnp1: 34 to 42 inches. Yellowish-brown (10YR-5/8) sandy clay loam with common medium-distinct mottles of red (2.5YR-4/8) and yellowish red (5YR-5/8); moderate, medium subangular blocky structure; firm, sticky; few soft and hard iron pebbles; soft plinthite; very strongly acid; gradual wavy boundary.

B24tp1: 42 to 60 inches. Reticulately mottled yellowish-brown (10YR-5/8), red (2.5YR-4/8), yellowish-red (5YR-5/8), and light-gray (10YR-7/1) sandy clay loam; moderate, medium subangular blocky structure; few patchy clay films on ped faces; firm, sticky; soft plinthite; very strongly acid.

Remarks: Colors are given for moist soil. Reaction determined by Soiltex.

WEIGHT PERCENT AND VOLUME PERCENT OF WATER RETAINED

DEPTH (inches)	TENSIONS (BARS)					BD G/CC	TP PCT	K
	.1	.3	.6	3.	15.			
0-11	10.81	6.84	5.77	4.16	2.07	1.41[1]	46.79	2.00 – 6.30
	15.24	9.64	8.14	5.87	2.92	1.49	43.77	
	FRAGMENT	7.87		SIEVED	1.98	ROCK PCT	38.74	
11-17	12.66	8.05	7.57	6.64	2.86	1.57[1]	40.75	0.63 – 2.00
	19.88	12.64	11.88	10.42	4.49	1.60	39.62	
	FRAGMENT	8.92		SIEVED	3.03	ROCK PCT	76.77	
17-24	16.14	14.23	10.77	10.52	6.13	1.33[1]	49.81	0.63 – 2.00
	21.47	18.93	14.32	13.99	8.15	1.33	49.81	
	FRAGMENT	13.70		SIEVED	5.08	ROCK PCT	6.70	
34-42	19.37	17.37	12.65	12.50	6.40	1.48[1]	44.15	0.63 – 2.00
	28.67	25.71	18.72	18.50	9.47	1.53	42.26	
	FRAGMENT	18.95		SIEVED	5.97	ROCK PCT	14.11	
42+	18.07	15.94	13.48	10.17	5.47	1.68[1]	36.60	0.63 – 2.00
	30.36	26.78	22.65	17.09	9.19	1.68	36.60	
	FRAGMENT	16.18		SIEVED	4.36	ROCK PCT	21.51	

1=FIST
2=CORE
3=LOOSE

SOIL TYPE - TIFTON LOAMY SAND
IDENTIFICATION CODE - 13122D
COVER - WEEDS-90, BARE-10
DATE OF RUN - 10 16 69
RAINFALL INTENSITY - 6.129 INCHES/HOUR
INITIAL SOIL MOISTURE FOR THE 0 TO 12 INCH DEPTH - 1.58 INCHES
INITIAL SOIL MOISTURE FOR THE 12 TO 36 INCH DEPTH - 6.14 INCHES
FINAL SOIL MOISTURE FOR THE 0 TO 12 INCH DEPTH - 2.92 INCHES
FINAL SOIL MOISTURE FOR THE 12 TO 36 INCH DEPTH - 7.39 INCHES

TIME FROM START OF RAIN (MINUTES)	ACCUMULATED RUNOFF (INCHES)	RUNOFF RATE (IN/HR)	ACCUMULATED INFILTRATION (INCHES)	INFILTRATION RATE (IN/HR)
7	0.000	0.000	0.715	6.129
10	0.001	0.060	1.018	6.069
15	0.040	0.881	1.492	5.247
20	0.100	0.952	1.942	5.176
25	0.197	1.258	2.357	4.871
30	0.303	1.360	2.760	4.769
35	0.423	1.511	3.152	4.618
40	0.561	1.777	3.525	4.352
45	0.722	2.068	3.874	4.060
50	0.902	2.260	4.205	3.869
55	1.094	2.365	4.524	3.764
60	1.293	2.477	4.835	3.652
65	1.506	2.560	5.133	3.568
70	1.722	2.606	5.428	3.523
75	1.931	2.556	5.730	3.572
80	2.144	2.560	6.028	3.568
85	2.361	2.609	6.321	3.520
90	2.565	2.535	6.628	3.593
95	2.777	2.534	6.927	3.595
100	2.988	2.491	7.227	3.638
105	3.202	2.477	7.524	3.651
110	3.416	2.497	7.821	3.632
115	3.623	2.439	8.125	3.690
120	3.841	2.551	8.417	3.577

SOIL TYPE — TIFTON LOAMY SAND
IDENTIFICATION CODE — 13122W
COVER — WEEDS-90, BARE-10
DATE OF RUN — 10 16 69
RAINFALL INTENSITY — 4.567 INCHES/HOUR
INITIAL SOIL MOISTURE FOR THE 0 TO 12 INCH DEPTH — 2.60 INCHES
INITIAL SOIL MOISTURE FOR THE 12 TO 36 INCH DEPTH — 6.90 INCHES
FINAL SOIL MOISTURE FOR THE 0 TO 12 INCH DEPTH — 2.97 INCHES
FINAL SOIL MOISTURE FOR THE 12 TO 36 INCH DEPTH — 7.34 INCHES

TIME FROM START OF RAIN (MINUTES)	ACCUMULATED RUNOFF (INCHES)	RUNOFF RATE (IN/HR)	ACCUMULATED INFILTRATION (INCHES)	INFILTRATION RATE (IN/HR)
5	0.000	0.000	0.380	4.567
10	0.075	1.605	0.686	2.961
15	0.200	1.670	0.941	2.897
20	0.360	2.202	1.161	2.364
25	0.560	2.438	1.342	2.128
30	0.760	2.427	1.522	2.140
35	0.967	2.477	1.696	2.089
40	1.171	2.467	1.873	2.099
45	1.372	2.427	2.053	2.139
50	1.577	2.447	2.228	2.120
55	1.782	2.403	2.403	2.163
60	1.977	2.328	2.590	2.238
65	2.176	2.377	2.771	2.189
70	2.373	2.336	2.955	2.230
75	2.563	2.295	3.145	2.271
80	2.759	2.303	3.330	2.264
85	2.936	2.173	3.533	2.393
90	3.120	2.089	3.730	2.478
95	3.299	2.072	3.931	2.494
100	3.472	2.008	4.139	2.558
105	3.657	2.125	4.334	2.441
110	3.832	2.123	4.540	2.443
115	3.998	2.002	4.755	2.564
120	4.184	2.119	4.950	2.448
125	4.356	2.052	5.158	2.514

TIFTON LOAMY SAND (14)

Location: 1.5 mi north of Coastal Plain Experiment Station dairy barn along station field roads; west for 350 yd along field road; 30 ft south of road; Tift County, Ga.

Land use or cover: Corn.

Great soil group: Plinthic paleudults; fine-loamy, siliceous, thermic.

Parent material: Unconsolidated marine sediments of sandy clay loam.

Drainage: Well drained.

Horizon and Description

Apcn: 0 to 10 inches. Dark grayish-brown (10YR–4/2) loamy sand; weak, fine. granular structure; very friable, nonsticky; many small hard iron pebbles one-eighth to one-half inch in diameter; many fine roots; strongly acid; abrupt smooth boundary.

B1tcn: 10 to 16 inches. Yellowish-brown (10YR–5/6) sandy loam; weak, medium granular structure; very friable, nonsticky; many small hard iron pebbles; fine roots common; strongly acid; clear wavy boundary.

B21tcn: 16 to 42 inches. Yellowish-brown (10YR–5/8) sandy clay loam; moderate, medium subangular blocky structure; friable, sticky; many small hard iron pebbles few fine roots mostly in the upper part; very strongly acid; gradual wavy boundary.

B22tcnp1: 42 to 52 inches. Yellowish-brown (10YR–5/6) sandy clay loam with common medium-distinct and prominent mottles of yellowish red (5YR–5/8), light yellowish brown (2.5YR–6/4), and red (2.5YR–4/8) moderate, medium subangular blocky structure; sticky; few soft and hard iron pebbles; soft plinthite very strongly acid; gradual wavy boundary.

B23tp1: 52 to 66 inches. Reticulately mottled yellowish-brown (10YR–5/6), light-gray (10YR–7/1), red (2.5YR–4/8), and light yellowish-brown (2.5YR–6/4) sandy clay loam; moderate, medium subangular blocky structure; few patchy clay films on ped faces; firm sticky; soft plinthite 10% to 30% by volume; very strongly acid.

Remarks: Colors are given for moist soil. Reaction determined by Soiltex.

TIFTON LOAMY SAND (14)

WEIGHT PERCENT AND VOLUME PERCENT OF WATER RETAINED

DEPTH (inches)	.1	.3	.6	TENSIONS (BARS) 3.	15.	BD G/CC	TP PCT	K
0-10	9.92	5.26	3.66	3.47	1.86	1.53[1]	42.26	2.00-6.30
	15.18	8.05	5.60	5.31	2.85	1.52	42.64	
	FRAGMENT	6.29		SIEVED	2.06	ROCK PCT	14.10	
10-16	10.95	5.45	5.14	3.69	2.74	1.58[1]	40.38	2.00-6.30
	17.30	8.61	8.12	5.83	4.33	1.61	39.25	
	FRAGMENT	4.48		SIEVED	2.81	ROCK PCT	19.45	
16-42	19.99	12.59	10.97	9.52	6.24	1.56[1]	41.13	0.63-2.00
	31.18	19.64	17.11	14.85	9.73	1.53	42.26	
	FRAGMENT	13.81		SIEVED	7.03	ROCK PCT	30.05	
42-52	22.56	13.49	11.42	9.13	2.40	1.58[1]	40.38	0.63-2.00
	35.64	21.31	18.04	14.43	3.79	1.56	41.13	
	FRAGMENT	15.89		SIEVED	2.77	ROCK PCT	11.25	
52+	17.16	13.02	12.48	9.34	2.54	1.63[1]	38.49	0.63-2.00
	27.97	21.22	20.34	15.22	4.14	1.67	36.98	
	FRAGMENT	13.30		SIEVED	1.24	ROCK PCT	3.01	

1=FIST
2=CORE
3=LOOSE

SOIL TYPE - TIFTON LOAMY SAND
IDENTIFICATION CODE - 14121D
COVER - WEEDS-90, BARE-10
DATE OF RUN - 11 10 69
RAINFALL INTENSITY - 4.687 INCHES/HOUR
INITIAL SOIL MOISTURE FOR THE 0 TO 12 INCH DEPTH - 1.94 INCHES
INITIAL SOIL MOISTURE FOR THE 12 TO 36 INCH DEPTH - 6.38 INCHES
FINAL SOIL MOISTURE FOR THE 0 TO 12 INCH DEPTH - 3.35 INCHES
FINAL SOIL MOISTURE FOR THE 12 TO 36 INCH DEPTH - 7.34 INCHES

TIME FROM START OF RAIN (MINUTES)	ACCUMULATED RUNOFF (INCHES)	RUNOFF RATE (IN/HR)	ACCUMULATED INFILTRATION (INCHES)	INFILTRATION RATE (IN/HR)
6	0.000	0.000	0.468	4.687
10	0.040	0.732	0.741	3.955
15	0.107	0.960	1.064	3.726
20	0.200	1.151	1.362	3.535
25	0.304	1.360	1.649	3.326
30	0.423	1.402	1.920	3.285
35	0.540	1.474	2.194	3.213
40	0.675	1.643	2.448	3.043
45	0.810	1.739	2.704	2.948
50	0.961	1.914	2.944	2.772
55	1.130	2.165	3.166	2.522
60	1.322	2.424	3.364	2.262
65	1.531	2.542	3.546	2.144
70	1.742	2.593	3.726	2.093
75	1.963	2.665	3.895	2.022
80	2.183	2.625	4.066	2.061
85	2.409	2.697	4.230	1.989
90	2.621	2.605	4.409	2.082
95	2.845	2.689	4.576	1.997
100	3.067	2.752	4.744	1.934
105	3.312	2.902	4.890	1.785
110	3.555	2.897	5.037	1.789
115	3.803	2.910	5.180	1.777
120	4.045	2.862	5.329	1.824

```
SOIL TYPE - TIFTON LOAMY SAND
IDENTIFICATION CODE - 14121W
COVER - WEEDS-90, BARE-10
DATE OF RUN - 11 10 69
RAINFALL INTENSITY - 6.370 INCHES/HOUR
INITIAL SOIL MOISTURE FOR THE 0 TO 12 INCH DEPTH - 2.97 INCHES
INITIAL SOIL MOISTURE FOR THE 12 TO 36 INCH DEPTH - 7.22 INCHES
FINAL SOIL MOISTURE FOR THE 0 TO 12 INCH DEPTH -  2.93 INCHES
FINAL SOIL MOISTURE FOR THE 12 TO 36 INCH DEPTH - -2.93 INCHES
```

TIME FROM START OF RAIN (MINUTES)	ACCUMULATED RUNOFF (INCHES)	RUNOFF RATE (IN/HR)	ACCUMULATED INFILTRATION (INCHES)	INFILTRATION RATE (IN/HR)
4	0.000	0.000	0.424	6.370
5	0.052	3.605	0.478	2.764
10	0.440	4.582	0.621	1.787
15	0.882	5.468	0.710	0.901
20	1.344	5.508	0.779	0.862
25	1.796	5.530	0.857	0.839
30	2.262	5.605	0.922	0.764
35	2.722	5.607	0.993	0.762
40	3.196	5.752	1.049	0.617
45	3.685	5.912	1.091	0.457
50	4.183	5.928	1.124	0.441
55	4.685	6.088	1.153	0.281
60	5.208	6.270	1.161	0.100
65	5.729	6.251	1.171	0.118
70	6.249	6.250	1.181	0.119
75	6.770	6.247	1.192	0.122
80	7.291	6.249	1.201	0.121
85	7.812	6.248	1.212	0.122
90	8.332	6.248	1.221	0.121
95	8.854	6.252	1.231	0.117
100	9.374	6.253	1.241	0.116
105	9.895	6.250	1.251	0.119
110	10.415	6.244	1.262	0.125
115	10.938	6.260	1.271	0.109
120	11.458	6.251	1.281	0.118

SOIL TYPE — TIFTON LOAMY SAND
IDENTIFICATION CODE — 14122W
COVER — WEEDS-90, BARE-10
DATE OF RUN — 11 13 69
RAINFALL INTENSITY — 2.743 INCHES/HOUR
INITIAL SOIL MOISTURE FOR THE 0 TO 12 INCH DEPTH — 3.06 INCHES
INITIAL SOIL MOISTURE FOR THE 12 TO 36 INCH DEPTH — 7.35 INCHES
FINAL SOIL MOISTURE FOR THE 0 TO 12 INCH DEPTH — 3.11 INCHES
FINAL SOIL MOISTURE FOR THE 12 TO 36 INCH DEPTH — 7.36 INCHES

TIME FROM START OF RAIN (MINUTES)	ACCUMULATED RUNOFF (INCHES)	RUNOFF RATE (IN/HR)	ACCUMULATED INFILTRATION (INCHES)	INFILTRATION RATE (IN/HR)
8	0.000	0.000	0.365	2.743
10	0.012	0.961	0.429	1.781
15	0.142	1.682	0.543	1.060
20	0.280	1.736	0.634	1.006
25	0.427	1.816	0.715	0.926
30	0.585	1.913	0.786	0.829
35	0.746	1.943	0.853	0.800
40	0.906	1.960	0.922	0.782
45	1.071	1.998	0.985	0.744
50	1.234	1.970	1.051	0.772
55	1.397	2.006	1.116	0.736
60	1.570	2.095	1.172	0.647
65	1.747	2.127	1.223	0.615
70	1.921	2.101	1.278	0.641
75	2.099	2.113	1.329	0.629
80	2.276	2.128	1.380	0.614
85	2.448	2.085	1.437	0.657
90	2.628	2.120	1.486	0.623
95	2.803	2.133	1.539	0.609
100	2.978	2.100	1.593	0.642
105	3.154	2.035	1.645	0.707
110	3.346	2.152	1.682	0.590
115	3.524	2.132	1.733	0.610
120	3.701	2.073	1.784	0.669

TIFTON LOAMY SAND (15)

Location: 1.5 mi north of Coastal Plain Experiment
Station dairy barn along station and field
road; west along field road for 0.5 mi; 15 ft
south of road; Tift County, Ga.

Land use or cover: Corn.

Topography: Very gently sloping — 3%.

Great soil group: Plinthic paleudults; fine-loamy, siliceous, thermic.

Parent material: Unconsolidated marine sediments of
sandy clay loam.

Drainage: Well drained.

Horizon and Description

Apcn: 0 to 6 inches. Dark grayish-brown (10YR-4/2)
loamy sand; weak, fine granular structure; very friable,
nonsticky; many small hard iron pebbles one-eighth to
one-half inch in diameter; many fine roots; strongly
acid; abrupt smooth boundary.

B21tcn: 6 to 32 inches. Strong brown (7.5YR-5/6)
sandy clay loam; moderate, medium subangular blocky
structure; friable, sticky; small hard iron pebbles common; few fine roots mostly in upper part; very strongly
acid; clear wavy boundary.

B22tcn: 32 to 46 inches. Brownish-yellow (10YR-6/6)
sandy clay loam with common medium and coarse mottles
of strong brown (7.5YR-5/8) and yellowish red (5YR-
5/8); moderate, medium subangular blocky structure;
friable, sticky; small hard pebbles common and few soft
iron pebbles; very strongly acid; gradual wavy boundary.

B23tp1: 46 to 64 inches. Reticulately mottled brownish-yellow (10YR-6/6), yellowish-red (5YR-5/8), red
(2.5YR-4/8), and light-gray (10YR-7/1) sandy clay
loam; moderate, medium subangular blocky structure;
patchy clay films on ped faces; firm, sticky; few soft
iron pebbles; soft plinthite 10% to 20% by volume; very
strongly acid.

Remarks: Colors are given for moist soil. Reaction determined by Soiltex.

TIFTON LOAMY SAND (15)

WEIGHT PERCENT AND VOLUME PERCENT OF WATER RETAINED

DEPTH (inches)	TENSIONS (BARS)					BD G/CC	TP PCT	K
	.1	.3	.6	3.	15.			
0-6	8.44	6.03	3.68	2.09	-0.11	1.56[1]	41.13	2.00-6.30
	13.17	9.41	5.74	3.26	-0.17	1.49	43.77	
	FRAGMENT	5.32		SIEVED	1.40	ROCK PCT	11.96	
6-32	21.56	13.98	12.40	6.81	3.02	1.58[1]	40.38	0.63-2.00
	34.06	22.09	19.59	10.76	4.77	1.61	39.25	
	FRAGMENT	13.68		SIEVED	4.82	ROCK PCT	9.00	
32-42	17.49	14.44	11.45	2.06	1.60	1.71[1]	35.47	0.63-2.00
	29.91	24.69	19.58	3.52	2.74	1.75	33.96	
	FRAGMENT	14.70		SIEVED	1.57	ROCK PCT	9.12	
42+	12.32	11.50	11.89	11.07	1.96	1.68[1]	36.60	0.63-2.00
	20.70	19.32	19.98	18.60	3.29	1.46	44.91	
	FRAGMENT	10.99		SIEVED	1.14	ROCK PCT	9.50	

1=FIST
2=CORE
3=LOOSE

SOIL TYPE - TIFTON LOAMY SAND
IDENTIFICATION CODE - 15121D
COVER - WEEDS-90, BARE-10
DATE OF RUN - 11 06 69
RAINFALL INTENSITY - 4.687 INCHES/HOUR
INITIAL SOIL MOISTURE FOR THE 0 TO 12 INCH DEPTH - 1.75 INCHES
INITIAL SOIL MOISTURE FOR THE 12 TO 36 INCH DEPTH - 6.52 INCHES
FINAL SOIL MOISTURE FOR THE 0 TO 12 INCH DEPTH - 3.43 INCHES
FINAL SOIL MOISTURE FOR THE 12 TO 36 INCH DEPTH - 7.28 INCHES

TIME FROM START OF RAIN (MINUTES)	ACCUMULATED RUNOFF (INCHES)	RUNOFF RATE (IN/HR)	ACCUMULATED INFILTRATION (INCHES)	INFILTRATION RATE (IN/HR)
4	0.000	0.000	0.312	4.687
5	0.012	0.240	0.378	4.446
10	0.024	0.231	0.757	4.456
15	0.052	0.404	1.119	4.282
20	0.092	0.553	1.470	4.134
25	0.144	0.710	1.808	3.977
30	0.208	0.770	2.135	3.916
35	0.272	0.805	2.462	3.882
40	0.343	0.892	2.781	3.794
45	0.424	1.046	3.091	3.641
50	0.519	1.252	3.386	3.434
55	0.634	1.372	3.662	3.314
60	0.746	1.377	3.941	3.310
65	0.863	1.423	4.214	3.263
70	0.982	1.577	4.486	3.109
75	1.121	1.691	4.737	2.995
80	1.260	1.725	4.988	2.962
85	1.413	1.920	5.226	2.766
90	1.582	2.071	5.448	2.616
95	1.760	2.200	5.660	2.486
100	1.941	2.223	5.870	2.463
105	2.131	2.353	6.070	2.333
110	2.331	2.396	6.262	2.291
115	2.532	2.397	6.451	2.289
120	2.731	2.396	6.643	2.290

SOIL TYPE - TIFTON LOAMY SAND
IDENTIFICATION CODE - 15121W
COVER - WEEDS-90, BARE-10
DATE OF RUN - 11 06 69
RAINFALL INTENSITY - 6.370 INCHES/HOUR
INITIAL SOIL MOISTURE FOR THE 0 TO 12 INCH DEPTH - 3.29 INCHES
INITIAL SOIL MOISTURE FOR THE 12 TO 36 INCH DEPTH - 7.29 INCHES
FINAL SOIL MOISTURE FOR THE 0 TO 12 INCH DEPTH - 3.30 INCHES
FINAL SOIL MOISTURE FOR THE 12 TO 36 INCH DEPTH - 7.25 INCHES

TIME FROM START OF RAIN (MINUTES)	ACCUMULATED RUNOFF (INCHES)	RUNOFF RATE (IN/HR)	ACCUMULATED INFILTRATION (INCHES)	INFILTRATION RATE (IN/HR)
4	0.000	0.000	0.424	6.370
5	0.048	3.004	0.482	3.365
10	0.300	2.995	0.761	3.374
15	0.565	3.147	1.027	3.222
20	0.855	3.895	1.268	2.474
25	1.208	4.400	1.446	1.969
30	1.574	4.519	1.610	1.850
35	1.959	4.740	1.756	1.630
40	2.358	4.922	1.887	1.447
45	2.776	5.049	2.000	1.320
50	3.200	5.101	2.108	1.268
55	3.645	5.288	2.193	1.081
60	4.079	5.248	2.290	1.121
65	4.523	5.318	2.376	1.052
70	4.959	5.297	2.471	1.072
75	5.400	5.225	2.562	1.144
80	5.864	5.327	2.628	1.043
85	6.339	5.627	2.684	0.743
90	6.762	5.342	2.792	1.027
95	7.229	5.543	2.856	0.826
100	7.687	5.585	2.929	0.784
105	8.135	5.486	3.012	0.883
110	8.593	5.488	3.085	0.881
115	9.066	5.657	3.142	0.712
120	9.500	5.431	3.239	0.938

SOIL TYPE - TIFTON LOAMY SAND
IDENTIFICATION CODE - 15122D
COVER - WEEDS-90, BARE-10
DATE OF RUN - 11 07 69
RAINFALL INTENSITY - 4.807 INCHES/HOUR
INITIAL SOIL MOISTURE FOR THE 0 TO 12 INCH DEPTH - 1.54 INCHES
INITIAL SOIL MOISTURE FOR THE 12 TO 36 INCH DEPTH - 7.54 INCHES
FINAL SOIL MOISTURE FOR THE 0 TO 12 INCH DEPTH - 4.06 INCHES
FINAL SOIL MOISTURE FOR THE 12 TO 36 INCH DEPTH - 9.55 INCHES

TIME FROM START OF RAIN (MINUTES)	ACCUMULATED RUNOFF (INCHES)	RUNOFF RATE (IN/HR)	ACCUMULATED INFILTRATION (INCHES)	INFILTRATION RATE (IN/HR)
4	0.000	0.000	0.320	4.807
5	0.008	0.600	0.392	4.206
10	0.060	0.592	0.741	4.215
15	0.116	0.677	1.085	4.129
20	0.172	0.697	1.430	4.109
25	0.231	0.714	1.771	4.092
30	0.292	0.720	2.111	4.087
35	0.351	0.702	2.453	4.105
40	0.413	0.750	2.792	4.057
45	0.477	0.776	3.128	4.031
50	0.541	0.767	3.465	4.040
55	0.606	0.801	3.800	4.005
60	0.675	0.868	4.132	3.939
65	0.746	0.904	4.461	3.902
70	0.822	0.944	4.786	3.863
75	0.902	1.024	5.106	3.783
80	0.989	1.085	5.420	3.722
85	1.082	1.166	5.728	3.641
90	1.182	1.214	6.029	3.593
95	1.281	1.193	6.330	3.614
100	1.381	1.194	6.630	3.612
105	1.481	1.189	6.931	3.617
110	1.582	1.195	7.231	3.612
115	1.684	1.227	7.530	3.580
120	1.781	1.186	7.833	3.621

SOIL TYPE - TIFTON LOAMY SAND
IDENTIFICATION CODE - 15122W
COVER - WEEDS-90, BARE-10
DATE OF RUN - 11 07 69
RAINFALL INTENSITY - 2.644 INCHES/HOUR
INITIAL SOIL MOISTURE FOR THE 0 TO 12 INCH DEPTH - 3.51 INCHES
INITIAL SOIL MOISTURE FOR THE 12 TO 36 INCH DEPTH - 9.31 INCHES
FINAL SOIL MOISTURE FOR THE 0 TO 12 INCH DEPTH - 4.07 INCHES
FINAL SOIL MOISTURE FOR THE 12 TO 36 INCH DEPTH - 9.54 INCHES

TIME FROM START OF RAIN (MINUTES)	ACCUMULATED RUNOFF (INCHES)	RUNOFF RATE (IN/HR)	ACCUMULATED INFILTRATION (INCHES)	INFILTRATION RATE (IN/HR)
5	0.000	0.000	0.220	2.644
10	0.020	0.238	0.420	2.405
15	0.040	0.191	0.620	2.452
20	0.096	0.757	0.785	1.887
25	0.160	0.775	0.941	1.868
30	0.224	0.767	1.097	1.876
35	0.288	0.756	1.254	1.887
40	0.351	0.795	1.411	1.849
45	0.424	0.921	1.558	1.722
50	0.505	0.964	1.698	1.679
55	0.585	1.008	1.838	1.636
60	0.674	1.105	1.969	1.538
65	0.766	1.150	2.097	1.493
70	0.861	1.167	2.223	1.476
75	0.959	1.154	2.346	1.489
80	1.048	1.066	2.477	1.577
85	1.137	1.019	2.608	1.624
90	1.221	1.035	2.744	1.608
95	1.309	0.989	2.876	1.654
100	1.384	0.835	3.021	1.808
105	1.453	0.798	3.173	1.845
110	1.521	0.796	3.326	1.847
115	1.590	0.817	3.477	1.826
120	1.658	0.816	3.629	1.827